Figure 3.2 (see Section 3.3, p. 7)

Mimicry in African butterflies.

Distasteful danaids

Female *Papilio dardanus* mimics

Male *Papilio dardanus*

Female *Papilio dardanus* (non-mimic)

Evolution

Second edition

Evolution

Second edition

Colin Patterson

Comstock Publishing Associates a division of
Cornell University Press   Ithaca, New York

First edition © Trustees of the British Museum (Natural History), 1978
Second edition © The Natural History Museum, London, 1999

The right of the author to be identified as the author of this work has
been asserted by the author in accordance with the Copyright Designs
and Patents Act of the United Kingdom, 1988.

First published in 1978 by Cornell University Press.
Second edition published in 1999 by Cornell University Press.

Every reasonable effort has been made to contact copyright holders and
to secure permission. In instances where this has not proved possible we
offer apologies to all concerned.

Printed in Belgium

Library of Congress Cataloguing-in-Publication Data #77-7865 is established for the
first edition. Second edition ISBN cloth 0-8014-3642-7 --- ISBN paperback 0-8014-8594-0

Cloth printing   10 9 8 7 6 5 4 3 2 1
Paperback printing   10 9 8 7 6 5 4 3 2 1

Contents

Preface to the First Edition

In writing this book I tried to produce an account of modern evolutionary theory which does not beg too many questions, and is complete enough to be coherent but simple enough to be comprehensible to those with little or no technical knowledge of biology. The book may not be easy reading for such people, mainly because an understanding of evolution requires a knowledge of genetics, which is not an easy subject. Nevertheless, I hope that the subject is important enough for the effort to be worthwhile. I have found it impossible to avoid using technical words, except by clumsy circumlocutions that would make the book much longer. But I have used as few as possible, tried to define each one the first time it is used, and there is a glossary at the end.

Preface to the Second Edition

When I wrote the first edition of this book I had one overriding aim, to keep it short and simple. If I had an audience in mind, I suppose it was my elder daughter, then 15 or 16 years old; time slips by and that child is now over 30. The intervening years have certainly been the most dramatic and exhilarating in the entire history of evolutionary biology, with important developments in techniques of study, in disputes over many aspects of doctrine and philosophy, and with evolutionary theory being brought into the courtroom and the polling booth by creationists. Of course, my own beliefs and opinions have shifted and vacillated during those years, and in sitting down to begin a second edition of the book I had only two alternatives: to stick to my original aim of keeping it short and simple, or to write an entirely different book. I settled on the first of those, not just out of indolence, but as the work went on I found that a rather different book was nevertheless emerging.

I could say that the first edition of the book was written in ignorance and the second in knowledge, or I could as well say that the first came out of knowledge and this one out of ignorance. I am not merely playing at paradoxes here. There is a knowledge, a certainty, that comes from education and indoctrination, and a knowledge or certainty that comes from working things out for yourself. Certainty and truth may or may not coincide. The knowledge in my first edition came from education and indoctrination; it was that neo-Darwinism is certainty. The knowledge in this second edition comes more from working things out for myself; it is that evolution is certainty. And part of the ignorance in the first edition concerned the difference between neo-Darwinism and evolution, whereas the ignorance in this edition is of the completeness of neo-Darwinism as an explanation of evolution.

To learn the differences between neo-Darwinism and evolution you may prefer to read on, but evolution is about what Darwin called 'descent with modification' – it concerns the idea of common or shared ancestry and the belief that all species are related by descent. I think that belief is now confirmed as completely as anything can be in the historical sciences. Neo-Darwinism concerns the explanation of descent with modification, and it emphasizes Darwin's own main contribution, natural selection. I am no longer certain that natural selection is the complete explanation, and I hope the new edition contains enough information, and not too much bias, for readers to understand the problem and judge the answer for themselves.

Biography

Colin Patterson was educated at Tonbridge School and Imperial College, London, where he took a first in Zoology in 1957. He began his career as a lecturer at Guy's Hospital Medical School, at the same time completing a PhD at University College. In 1962 he was appointed to the Palaeontology Department of The Natural History Museum and remained at the Museum – apart from a spell at Harvard University in 1970 – until his death in March 1998. Though retired in 1993 at the age of 60 he continued to work daily at the Museum.

Colin's publications began in 1963, and numbered more than 100, the majority of them papers and monographs on the structure and evolution of fishes. His many honours included his election as a Fellow of the Royal Society in 1993, Research Associate at the American Museum of Natural History, member of the Board of the National Museum of Natural History, Smithsonian Institution and Fellow of the Linnean Society.

Foreword

Colin Patterson died just three days after delivering the final version of the manuscript of this book leaving the final checking of the manuscript and artwork unfinished. We have been privileged to take up that task but also slightly daunted since Colin's ability at such tasks was unrivalled. We have left the book virtually unchanged, merely correcting minor errors of fact and checking artwork, final versions of which Colin never saw.

Working alongside Colin for 23 years one of us (PLF) has had an opportunity to see at first-hand his intellectual sparring with the Theory of Evolution by Natural Selection. Colin entered the field of vertebrate palaeontology convinced that it was there that the broad features of evolution would be revealed through patterns of ancestry and descent entombed in rock; such was the paradigm in the 1950s. He quickly came to realise that the fossil record was mute on matters of process but he maintained his interest in the Theory about which he read copiously and critically, as he did on every subject. His favourite critical internalised question was 'how do we know that?': to which he often got the answer 'authority or tradition' — something which he respected only after he had explored the evidence for himself. For instance, he asks after the raw data on which the changes in gene frequency in the Scarlet Tiger Moth are used by proponents of two sides of an argument concerning natural selection, and finds it unconvincing for either's case. This book is written very much in this investigative mode, and it is done in his personal, clear and near conversational style which appeals to many levels of readership. The science is blended seamlessly with anecdotes about the cladistics battles in The Natural History Museum, and quotes from Little Richard and Margaret Thatcher.

Beneath that smooth and comforting style there are some points that have become crystallised from liquid thoughts in the first edition and textbooks by other authors. For instance, Colin now considers that genetic variation may be the result of largely random processes and acknowledges that the theory of macromutation, rejected outright in the 1950s, has reappeared, although metamorphosed, especially in the duplication of homeobox genes to allow major morphological radiation.

As a lecturer in evolutionary genetics one of us (JM) is a connoisseur of evolution texts, which, like the organisms they treat, are often derivative of one another. This work is a refreshing contrast. It is a pleasure to find this book so enjoyable, original and informative, allowing the reader a quick and understandable way into a field of scientific enquiry which underpins all of Biology.

Peter L. Forey
Department of Palaeontology
The Natural History Museum
Cromwell Road
London SW7 5BD

James Mallet
Reader in Evolutionary Biology
University College London
4 Stephenson Way
London NW1 2HE

Propositions

The theory of evolution is the foundation of contemporary biological science. It unifies and directs work in all sorts of specialized fields, from medicine to geology. The modern theory is often called neo-Darwinian: 'Darwinian' because it uses Darwin's idea of natural selection, and 'neo' (new) because it incorporates a theory of heredity, worked out since Darwin's time. Neo-Darwinian theory can be summarized in six propositions:

- Reproduction
  'Like begets like': reproduction in populations of organisms produces descendent populations of similar organisms.
- Excess
  The reproductive potential of the parent population always greatly exceeds the actual number of its descendants.
- Variation
  Members of a population always vary. Much of this variation is transmitted to the descendants (heritable), and novelties (mutations) may appear.
- Environmental selection
  The space and resources of the environment are limited, so that there is competition within and between populations. Individuals possessing favourable characteristics, of whatever sort, will tend to compete successfully and leave more descendants than other, less lucky individuals.

- Divergence
  The environment varies with time and from place to place. Heritable variations that suit a particular environment will be selected there, and so populations will diverge and differentiate as each becomes adapted to its own conditions. The terms 'adapted' and 'adaptation' have special meanings in biology. We commonly think of **adaptation** as part of a lifetime ('he soon adapted to city life'), but in Darwin's usage and in modern biology the term may mean a process that operates over many generations, or a feature thought to result from that process, such as a prehensile tail or a parrot's beak.
- Common ancestry
  The principle of divergence has no limit, and the diversity of life on Earth can be explained by divergent descent of lineages from more or less remote common ancestors.

This outline is amplified, explained and criticized in the rest of the book. Modern evolutionary theory is a large subject, difficult to set out as one connected argument. This is partly because nature is so diverse that exceptions can be found to almost every statement and partly because evolution involves contributions to and from every branch of biology. In 1859, Charles Darwin chose to call his book *The Origin of Species*, because the existence of species and their variety were the fundamentals he wanted to explain. Now, as then, the idea of species is a good place to begin the argument.

2 Species

The chicken is the egg's way of ensuring the production of another egg.
Samuel Butler

The concept of the **biological species** is central to evolutionary thought. **Species** (the plural is the same) is a Latin word meaning 'kind' or 'sort'. Among animals familiar to us, the distinction between a blackbird and a thrush or between a dog and a cat is simple enough. Before Darwin such distinctions were sufficiently explained by the notion that these animals belonged to different species. But what are we actually recognizing when we learn to call one bird a blackbird and another a thrush? Not simply that these are two kinds of animals that look different, because most people can distinguish male from female blackbirds, and a blackbird's nest from a thrush's nest by its form and by the eggs it contains. We also learn that one sort of caterpillar is a cabbage white, while another is a red admiral: surely a blackbird's egg 'is' a blackbird and an acorn 'is' an oak tree, in just the same way.

The relationship that links the acorn and the oak tree or the caterpillar and the butterfly gives a clearer picture of the biological species. It is a community of plants or animals in which each member is the offspring of two earlier members and is in turn capable of producing other members by mating within the community.

This idea of species – interbreeding communities extending in successive generations through time – can apply only to organisms with the two-parent pattern of descent with which we are familiar. Where an organism always reproduces asexually – for example by budding, or splitting, or self-fertilizing – the offspring of each parent will be isolated for ever from the rest of the community (see Box 2.2). We cannot apply a biological species concept (potential interbreeding community) to such organisms and must rely on similarities and differences to recognize species. Many species reproduce asexually but few are entirely without some sort of sexual process, and those that are not have much of a future (see Section 15.5).

The idea behind the biological species – community of descent rather than community of resemblance – can be applied to all kinds of organisms, including simple types that cannot easily be separated into species by obvious features like those that distinguish thrushes and blackbirds. The test, in doubtful cases, is ease of interbreeding and the capacity to produce healthy fertile offspring. Thus:

- All human races are fully interfertile and so belong to a single species.
- A cross between a male donkey and a female horse produces vigorous offspring, the mule, but all mules are sterile. No mixed race can result, and horses and donkeys are therefore different species.

On the other hand, the Chiffchaff and Willow Warbler (common English birds) are so similar that it was not until the late eighteenth century, when Gilbert White of Selborne noticed a difference in song, that they were

---

Box 2.1

Species names and common names

'Blackbird', 'song thrush' and 'mistle thrush' are the English names given to three biological species. These birds also live in many other countries, and have different names in the language of each. In biology each species is given a name consisting of two Latin or latinized words. The first, with an initial capital, is the **generic name** (of a genus), and may be shared by several species (see Section 13.1 on classification, page 100), and the second is the **specific name**. The blackbird is *Turdus merula*, the song thrush *Turdus philomelos* and the mistle thrush *Turdus viscivorus*. The advantage of these names is that a Russian- or Chinese-speaking biologist or bird watcher will understand them as precisely as one speaking English or French.

and larvae differ, and each species has its own geographic range, breeding and feeding behaviour, tolerance of salt water, and so on. The importance of this work to malaria control is that only three of the six species transmit malaria. From our point of view, the important thing is that these six species are real, non-interbreeding units, recognized only after long study: species are real, not a human concept that we try to impose on nature.

I do not pretend that the ability to interbreed is an infallible test, or a definition, of species. No one has yet succeeded in producing a definition of species that covers all cases. But definitions are usually only interesting to those who wish to quarrel with them. And if Darwin is right – that all species are the product of change, of divergence from common ancestors – we should expect to find examples of species in the making, which can be distinguished only arbitrarily or with difficulty. It remains true that species are real, distinct units in nature.

If the living world can be partitioned into species in this way, then the first questions we should ask are not whether such communities change through time (evolution) but why they remain distinct and how it is that perfect copies of the parents are produced, generation after generation. Such questions, which perplexed Darwin, have been answered by the study of heredity (discussed in Chapter 4).

But before starting on the intricacies of genetics, other questions need to be asked about species:

● Are they all of the same sort?

● What variations exist between members of one species?

suspected to be different. It is now known that they never interbreed in nature, and behave as separate species.

The mosquitoes that transmit malaria in Europe provide a more complicated example. Early attempts to control malaria by eliminating the mosquito met with little success. Intensive study showed that much of the anti-mosquito campaign had been wrongly directed. What at first seemed to be a single species of mosquito, *Anopheles maculipennis*, was resolved into six quite distinct species. The adults of these species are virtually indistinguishable, but the eggs

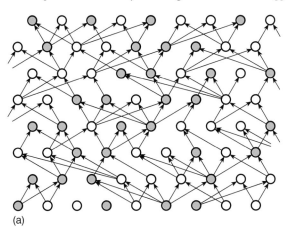

(a)   (b)

Figure 2.1

Population structure in sexual and asexual species

In a sexual species (a), two parents of different sexes are needed to produce one offspring but in an asexual species (b) each individual has only one parent.

3   Variation within species

We can speak of offspring as copies of their parents. In our own species, we know that this copying is not exact, otherwise we should have trouble recognizing our friends and families. Nevertheless, these imperfections in copying are usually in the finer details – shade of eyes and hair, or cast of features. If we imagine having to recognize our friends by their feet rather than their faces, we realize how finely tuned is our appreciation of human variation: presented with a brood of houseflies we would be ready to admit the perfection of the copying process, as would a member of another species looking at humans. But because our appreciation of variation is keenest within our own species, it is helpful to use humans as an example of the kinds of variation found in nature. Natural variation seems to fall into three categories:

- individual variation (each individual of a species is distinguishable from the others);
- geographic variation (individuals of a species in one district differ from others in another);
- polymorphism (when two or more well-marked varieties coexist in the same district).

The most conspicuous variation of all occurs not in nature but under the artificial conditions of domestication: this is a fourth category.

3.1   Individual variation

In the human species individual variation includes the sort of differences that enable us to recognize acquaintances and public figures – differences in features, stature, voice, etc. More precise indications of the uniqueness of each person come from fingerprints: the prints of only one or two fingers may be sufficient to trace an individual. 'Genetic fingerprinting' or 'DNA profiling' (Section 5.4), a technique that exploits molecular individuality, has now come into the courtroom alongside or instead of traditional fingerprinting.

Although this kind of analysis has not yet been taken as far in other species as it has in our own, it is evident that similar variations exist in most animals, so that individuals are unique in several recognizable ways.

3.2   Geographic variation

In general, geographic variation within a species is on a larger scale and involves more obvious differences than the details which distinguish individuals from the same region. In humans, the difference between the races – Africans, Caucasians, Orientals etc. – is primarily a geographic phenomenon. The link with geography is less obvious now, after centuries of (increasingly rapid) transportation, but the early navigators observed it without difficulty. The differences between human races are often more obvious than those between individuals of any one race, but they concern relatively few features – skin coloration, stature, hair form and colour, some facial characteristics, etc. And, although the differences between individuals of a single race may seem to be random and without function, some of the differences between human races can be seen as adaptations to climate or other conditions. For example, in the tropics dark skin is advantageous since it shields the underlying tissues from sunburn but in high latitudes, where sunburn is less of a danger, enough of the ultraviolet component of sunlight must reach the tissues to produce vitamin D (deficiency in vitamin D results in rickets, and the vitamin is synthesized with the help of ultraviolet radiation) – so pale, translucent skin is an advantage.

We find just the same phenomena in other species: individuals of a species from one region may be almost indistinguishable, but those from widely separated regions can often be differentiated easily, and may differ in features which are adapted to local conditions.

Figure 3.1

Circumpolar ring species of gulls

Circumpolar ring species of Herring Gull (*L. argentatus*) and Lesser Black-backed Gull (*L. fuscus*). The hatched area is the overlap between the two species, and the arrows mark the division between them according to different authorities. Lesser Black-backed Gulls are darker and smaller in Norway and the Baltic (subspecies *fuscus*), paler and larger in Britain and Iceland (subspecies *graellsii*). Some authorities believe that Taimyr Gulls are hybrids between Vega and Heuglin's Gulls.

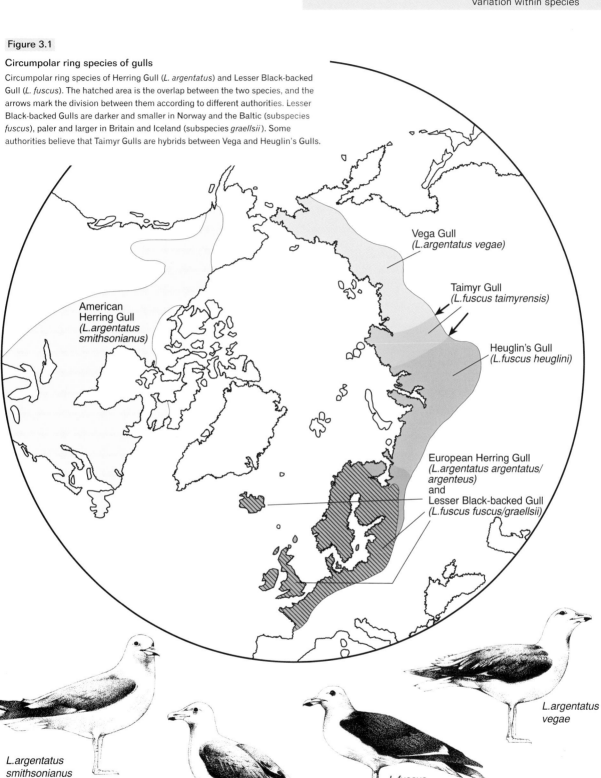

Vega Gull
(*L. argentatus vegae*)

Taimyr Gull
(*L. fuscus taimyrensis*)

Heuglin's Gull
(*L. fuscus heuglini*)

American
Herring Gull
(*L. argentatus
smithsonianus*)

European Herring Gull
(*L. argentatus argentatus/
argenteus*)
and
Lesser Black-backed Gull
(*L. fuscus fuscus/graellsii*)

*L. argentatus
smithsonianus*

*L. argentatus
argentatus*

*L. fuscus
fuscus/graellsii*

*L. argentatus
vegae*

Two common British birds, the Herring Gull (*Larus argentatus*) and the Lesser Black-backed Gull (*Larus fuscus*), are often seen together beside the Thames in London. In north-west Europe the populations of these two gulls are sharply distinct – in appearance, habits, migratory pattern and so on. Interbreeding sometimes produces fertile offspring, but it is very rare in the wild and the two birds behave as distinct biological species. Further east, in northern Europe and Asia, there is only one species. In north-west Russia the gulls have dark backs, although not so dark as in the Lesser Black-backed Gull, and yellow legs (unlike the pink of the British Herring Gull). In north-east Siberia the gulls have paler backs and pink legs. In North America the gulls have very pale backs and pink legs, and look much like European Herring Gulls. Lesser Black-backed Gulls did not occur west of the British Isles until recently, but they colonized Iceland in the 1920s and regular sightings are now made on the east coast of North America.

In north-west Europe there are clearly two distinct species (*L. argentatus*, *L. fuscus*), because they live together but remain separate. Across North America the gulls are very like European Herring Gulls, and no one doubts that they belong to *L. argentatus*. Vega Gulls, found in north-eastern Siberia, have darker backs than American Herring Gulls but no one doubts that they are closer to *L. argentatus* than to *L. fuscus*. The problems begin in north Russia and north-west Siberia: are the gulls *L. argentatus*, *L. fuscus*, or something in between? The authorities are divided.

One authority wrote recently 'As the study of gull identification becomes increasingly sophisticated, we are discovering that we actually know less about these birds than we thought we did.' But the birds themselves know better and presumably the authorities will eventually learn from them, as they have recently learned about Yellow-legged Gulls (Box 3.1).

This situation, where two apparently distinct species are joined by a series of geographical and structural intermediates, is called a ring-species. The sketch map in Fig. 3.1 shows the breeding range of **subspecies** of Herring Gulls (*L. argentatus*) and Lesser Black-backed Gulls (*L. fuscus*).

To the naturalist who wishes to place the creatures he studies in neat pigeonholes, ring-species are a problem. But they are welcome to the evolutionist, because if species change with time and may split into two, we should expect to find that some present-day species are not sharply distinct – some will be in the process of splitting and will present difficult decisions about the number of species. In the case of the gulls described above, current opinion is that the

---

Box 3.1

The Yellow-legged Gull

A southern series of subspecies, Yellow-legged Herring Gulls, extends from Spain, through the Mediterranean, Black and Caspian Seas, and into central Asia as far as Mongolia. Until very recently these southern gulls were included either in *L. argentatus* (by ornithologists who believed that they are closest to Herring Gulls) or in *L. fuscus* (by those who thought them closer to Lesser Black-backed Gulls). The problem of species has now been solved by the birds themselves. In this century, gull populations have increased enormously in Europe, probably as a result of human population growth (gulls thrive on rubbish dumps and on offal from the fishing industry). During the last 20 years, Herring Gulls and Lesser Black-backed Gulls have expanded southwards and Yellow-legged Gulls have expanded northwards. All three now breed together in France, but Yellow-legged Gulls have not yet interbred successfully with either Herring or Lesser Black-backed Gulls, demonstrating that the three are distinct species (Yellow-legged Gulls are now called *Larus cachinnans*).

---

two terminal, overlapping populations are different species and that the boundary between them lies somewhere in northern Russia. The various dark-backed populations west of this boundary are named as subspecies of *L. fuscus* and the light-backed populations to the east of the boundary as subspecies of *L. argentatus*. However, disagreement among the experts (as shown in Fig. 3.1) makes it obvious that these decisions are arbitrary.

Difficult or arbitrary decisions like this are common. In the British Isles (a series of outliers of continental Europe), many terrestrial European animals and plant species are represented in mainland Britain and in other British islands by populations that are separated from their neighbours by water barriers. These populations can often be distinguished by slight differences in proportions, size, habits and so on. Many species and subspecies of small mammals (such as voles and field mice) and small birds (like wrens) have been named in the various islands, especially in the Shetlands and Hebrides. The decision whether these are distinct biological species or geographical variants of the mainland form is arbitrary, because the test of interbreeding could only be made after capturing and transporting individuals, treatment that could well influence behaviour.

Geographic variation is widespread, and the idea of biological species cannot be applied inflexibly to every situation found in nature.

Continuous geographic variation

The examples of geographic variation just discussed are not obviously related to environmental differences. But some types of geographic variation are adaptive, and are so widespread that rules have been formulated to describe them. In warm-blooded animals (birds and mammals), members of a species living in high (cold) latitudes are generally larger than their relatives at low latitudes and have shorter ears, tails and limbs. The advantage of large size and short extremities in cold climates is that heat loss is reduced because the ratio of surface area (which radiates heat) to volume is lower in large animals than in small. In hot climates, long extremities and small size increase heat loss. Many plant species vary in size, growth rate, chlorophyll content and other features according to the altitude at which the population is growing. Continuous geographic variation of this sort, following latitude or altitude, may appear to be only the direct result of the conditions under which an individual develops but (as we shall see in Chapter 5) such variations are usually inherited.

3.3   Polymorphism

Polymorphism denotes a situation in which two or more clear-cut variants of a species coexist in the same region. A human example of polymorphism is the four major blood groups (A, B, AB and O). Every individual belongs to one of these groups, and no intermediates occur. Hidden polymorphisms like this, which are detectable only by biochemical tests, are common in animals and plants, but more attention has been paid to polymorphism in conspicuous features. Variations can be assumed to be inherited, not determined by the conditions of life, and they are particularly suitable for investigating the effects of natural selection. Examples include the snail *Cepaea nemoralis*, which is common in England and is polymorphic for shell colour and the presence or absence of dark bands around the shell (see Fig. 8.7), and the peppered moth, *Biston betularia*, which is polymorphic for wing colour, with pale and dark forms (see Fig. 8.3). In these examples the different variants have advantages under certain environmental conditions (see Chapter 8). The adaptive significance of the human ABO blood group system is not clear, although the incidence of some cancers is lower in individuals of group O than A, and AB individuals seem less susceptible to other diseases than those of blood group O.

One particularly intricate type of adaptive polymorphism is the mimicry found in various butterflies such as the African *Papilio dardanus* (Fig. 3.2, see frontispiece). The males of this species are all of one type (monomorphic), but the females occur in several forms that are very different from the males and from each other. Each of the conspicuous female variants resembles a different species of another group of butterflies, the danaids. These are protected from predators such as birds by being unpalatable. *Papilio dardanus* is good to eat, but the various female mimics are protected because birds that have learned not to eat the distasteful danaids will avoid them. If the mimics were as numerous as the danaids they copy, birds would be slow to learn to associate the wing pattern with unpleasant taste because they would occasionally eat a tasty *Papilio*. By imitating one of several different species, each female *Papilio dardanus* gains a greater advantage from mimicry than if all mimicked the same species.

3.4   Variation under domestication

Darwin used variation in domestic animals as one of the major arguments for his theory. The science of genetics was then unknown and, apart from human family trees, domestic animals and plants were the only source of reliable information about patterns of inheritance. As early as 1840 Darwin produced an elaborate questionnaire on animal breeding for circulation to stockbreeders and others. Darwin bred pigeons himself, and made good use of his knowledge of them in *The Origin of Species*. When John Murray, his publisher, submitted the manuscript of that book to the Reverend Whitwell Elwin, a respected scholar, for an opinion of its worth, Elwin suggested that Darwin confine himself to pigeons, writing 'Everybody is interested in pigeons. The book would be reviewed in every journal in the kingdom, and would soon be on every library table.' The second sentence, at least, was prophetic.

The most striking example of variation in domestic animals is surely the amazing range of dogs. Darwin thought that the various breeds of dog were descended from several different wild species. He gave no good reason for this opinion, and it seems that he was simply unable to believe that a single species could be the source of such variety.

Modern opinion is that the domestic dog is descended from the wolf, *Canis lupus* (Fig. 3.3). Archaeological finds prove that dogs have been domesticated for at least 14 000 years. We do not know whether wolves were domesticated once or several times, or where the original domestication took place, although it must have been in the northern hemisphere, the home of the wolf. Dogs have since been distributed all over the world by humans, and in many places they have escaped and assumed the status of a wild species, like the dingo in Australia.

The results of breeding and selecting dogs are too familiar to need description. All breeds of dog are interfertile, but some crosses, for example between a 1 kg Chihuahua and a 75 kg Great Dane, are prevented by the disparity in size. A cross between a Great Dane and a St Bernard results in seriously defective offspring, but this is not because the parents belong to different species. It is because each of these breeds has been selected for giant size as far as it will go, and crossing these two specialized lines of giants results in partial breakdown of the mechanism of development. In the same way, bulldogs have been selected for their peculiar characteristics to the point where many males are infertile.

We can take the many breeds of dog as an indication of the potential variability inherent in a single species of mammal. This variation is genetic, since it breeds true. In dogs, the range of form is very great because selection has been guided not only by the desire to fit form to function, as in working dogs, but also (in 'toy' breeds) by whim or fashion. In other domesticated mammals, such as horses, pigs and sheep, the variation produced by selective breeding from single wild species is striking (compare a racehorse with a Shetland pony), but not as great as in dogs. This may be because these animals have been bred for use, not caprice, and perhaps because they have not been domesticated as long as the dog (the earliest remains of domestic pigs and cattle are only about 8500 years old).

Darwin's account of pigeon breeds (often selected, like dogs, for fancy rather than use) shows that bird species have as much potential variability as mammals. The work of agriculturists with cereals and other vegetables, and of geneticists with insects, confirms that great potential variability is a feature of all species.

Figure 3.3

Variation in domestic animals

A wolf (*Canis lupus*) and an assortment of domestic dogs descended from the wild species.

3.5 Summary

This survey of species and their variation can be summed up as a series of statements, which in turn provoke questions to be answered by the study of heredity.

1. Definition of the biological species as an interbreeding community does not always work perfectly. In some cases, like ring-species, decisions about the limits of species are arbitrary.

- What mechanisms cause members of a species to develop as near-perfect copies of their parents?
- What prevents members of different species from interbreeding?
- Are there genetic mechanisms that allow some individuals to interbreed successfully and prevent others from doing so?

2. Within a species, each individual is usually unique, characterized by variations in structure and chemistry.

- Are these variations inherited?
- If so, what is their source?

3. Many species contain recognizable geographic races or subspecies, often adapted to local conditions.

- Are these adaptations inherited, or induced by the conditions?

4. Many species are polymorphic, with clear-cut variants living together in the same region.

- How are these variations inherited and maintained?

5. Domestication shows that wild species contain enormous potential heritable variation.

- What is the source of this variation, and why is it not manifested in the wild?

4   Heredity

The laws governing inheritance are quite unknown.
Charles Darwin, *The Origin of Species*

A great man of science … knows everything about everything, except why a hen's egg don't turn into a crocodile, and two or three other little things.
Charles Kingsley, *The Water Babies*

We recognize biological species as communities linked or held together by descent, by parentage. In most species that breed sexually the only material that passes from the parents to the next generation is the minute amount contained in the egg and sperm (the human egg, for example, is about 0.1 mm across and the sperm is very much smaller). The fertilized eggs of the vast majority of species are left to take care of themselves. Since such eggs develop into copies of the parents, they must contain instructions for doing so.

Darwin was completely ignorant of the way in which these instructions are passed from generation to generation, despite his experience in breeding animals and plants. He knew that inheritance is often 'blending' – the offspring being 'of mixed blood', intermediate between the parents – but he also knew of many exceptions to that rule. He believed that characteristics acquired during the life of an organism, such as parts or organs enlarged by use or weakened by disuse, were transmitted to the offspring, and in 1868 published a theory of heredity that seemed to explain this. He proposed that all parts of an organism secrete minute granules throughout life, which circulate around the body and accumulate in the reproductive organs, ready to be passed on to the next generation. However, two years before, in 1866, Gregor Mendel published an account of a long series of breeding experiments with pea plants, from which he developed a theory that explained many of Darwin's difficulties (and eventually showed Darwin's own

theory of heredity to be completely false). Unfortunately, the significance of Mendel's work was not noticed by Darwin – or any other scientist – until 1900, when both Mendel and Darwin were dead.

4.1   Mendel's theory

To proceed further with the problem 'What is a species?' as distinct from the other problem 'How do species survive?' we must go back and take up the thread of the enquiry exactly where Mendel dropped it.
William Bateson, 1901

Mendel experimented with varieties of garden peas (*Pisum sativum*). He found varieties with clear-cut differences in features such as flower colour (white versus purple), seed colour (green versus yellow) and form of seed (round versus wrinkled), and grew each variety for two years to make sure that it was pure – that it bred true. He chose pea plants because the flower is normally self-fertilized, but can be cross-fertilized experimentally if the stamens are removed before they ripen.

In a typical experiment (Fig. 4.1) Mendel cross-fertilized the flowers of plants grown from green and yellow seeds and found that the resulting hybrid seeds (peas) were all yellow. He then planted these hybrids, allowed them to self-fertilize and examined the seeds produced. He found that the hybrid plants, grown from yellow seed, did not breed true. Out of 8023 peas in this second generation, 6022 were yellow and 2001 green – a ratio very close to three yellow to one green. Again, he planted these peas and allowed the second-generation plants to self-fertilize. He found that the green peas always bred true, but that the yellow ones did not. Out of 519 second-generation yellow peas, 166 bred

---

Box 4.1

In a mature pea plant, the plant, including the pods, represents the parent generation. The flesh of the peas within the pod belongs to the next generation, but the seed coat, the translucent 'skin' of the pea, is parental tissue like the pod.

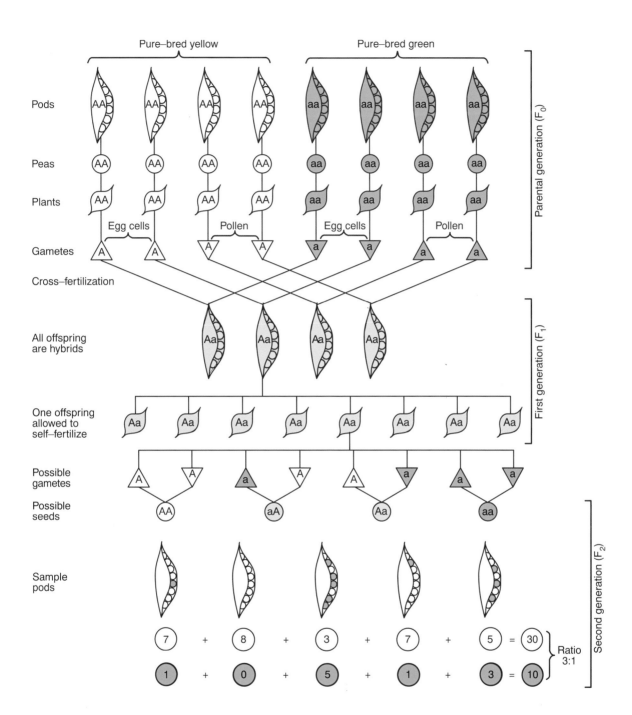

Figure 4.1

Mendel's experiments with green and yellow peas

Mendel's experiments. Cross-fertilizing pure-bred green and yellow peas produced all hybrid offspring in the first (F_1) generation. None of these hybrids bred true. The proportion of three yellow to one green peas in the second (F_2) generation is an ideal result and would be only approximated in actual experiments.

true but 353 did not: the latter group again produced yellow and green seeds in the ratio of three to one. When a plant yielded yellow and green seeds, the different types were randomly distributed in the pod: a pod might contain five or six yellow and two or three green peas.

Mendel's fame is due to the theory he devised to explain these facts. First, since a cross between the pure-bred yellow line and the pure-bred green produced all yellow peas, he supposed that the character 'yellow' was **dominant** (symbolized by 'A' in Fig. 4.1) over 'green', which he called **recessive** (symbolized by 'a'). Since a hybrid yellow pea can produce both green and yellow offspring, it must carry the 'green' character in latent form. To explain this, he proposed that each pea and plant carries a double dose of characters or hereditary factors (now called **genes**), so that pure-bred yellow could be symbolized by AA, pure-bred green by aa. The pollen and egg cells would carry only one factor, A or a. The first-generation peas in his experiment (F_1), all yellow, would be Aa, having received 'A' (in pollen or egg) from one parent and 'a' from the other. The second generation (F_2), produced by self-fertilization of the first, should then have an equal chance of receiving either 'A' or 'a' in both the pollen and the egg cell, and so should be 25% AA (pure-breeding yellow), 50% Aa (like their parents, impure breeding yellow) and 25% aa (pure-breeding green): thus the second generation should contain three-quarters yellow ($1/4$ AA + $1/2$ Aa) and one-quarter green (aa), which explains the 3:1 ratio that Mendel observed. By a series of crosses between plants of each constitution Mendel was able to test this theory, and to show that it accounted for all the available facts. He also found that reciprocal crosses (pollen from green crossed with egg from yellow; yellow pollen with green egg) proved that the allocation of the genetic factors (A or a) to the pollen and egg (male and female sex cells, or **gametes**) is random: each gamete has an equal chance of receiving either factor. Fertilization is also random, an egg-cell with A having an equal chance of being fertilized by pollen with A or a.

Most subsequent work in genetics has been based on Mendel's principles and practices. The great advances that Mendel made were:

- his ideas of dominant and recessive factors (genes);
- the double dose of factors in an organism, so that an organism can be hybrid;
- the random segregation of single factors to the gametes, so that a gamete is always pure, carrying one factor or the other; and

- the transmission of the factors, unchanged and uncontaminated, from generation to generation.

Mendel showed that inheritance is not blending, as Darwin supposed, but particulate (Mendel's particles, the genes, are nothing to do with the mythical 'granules' that Darwin supposed were given off by all parts of the organism).

One other point of great importance that follows from Mendel's theory is the clear distinction between the physical appearance of an organism (its **phenotype**) and its genetic constitution (**genotype**). The same phenotype may have different genotypes (e.g. yellow peas (phenotype) may have the genotype AA or Aa), or one genotype may have several phenotypes (e.g. the caterpillar, the chrysalis and the butterfly are three different phenotypes, all expressing the same genotype).

Mendel also conducted more complicated experiments involving two pairs of factors, for example crossing pea plants grown from pure-bred yellow, wrinkled seeds with plants grown from green, smooth seeds. He found that the yellow/green pair of factors and the wrinkled/smooth pair were inherited independently.

Subsequent experiments have confirmed all these aspects of Mendel's work, and shown that his principles can be applied in all sexual organisms, but we now know that clear-cut dominant/recessive systems involving a single pair of factors are not all that common – in many attributes, such as height in human beings, the offspring are roughly intermediate between their parents. But analysis shows that this apparent blending of parental features is not a different type of inheritance: it involves attributes that are controlled by several or many Mendelian genes, with additive effects. Mendel's rule that different pairs of factors are inherited independently also has many exceptions: factors tend to be transmitted in groups, called **linkage groups**. These, and other apparent anomalies, were first explained by microscopic study of the cells of organisms.

4.2 The chromosome theory

By 1900, when Mendel's work was rediscovered, a quite different line of investigation could be brought into genetic theory: direct observation of cells under the microscope.

All organisms are composed of one or more cells, and all plant and animal cells consist of two zones – an outer zone, the **cytoplasm**, containing the various organelles (microscopic organs) concerned with the life processes of the cell, and an inner zone, the **nucleus**, which is concerned with reproduction (among other things).

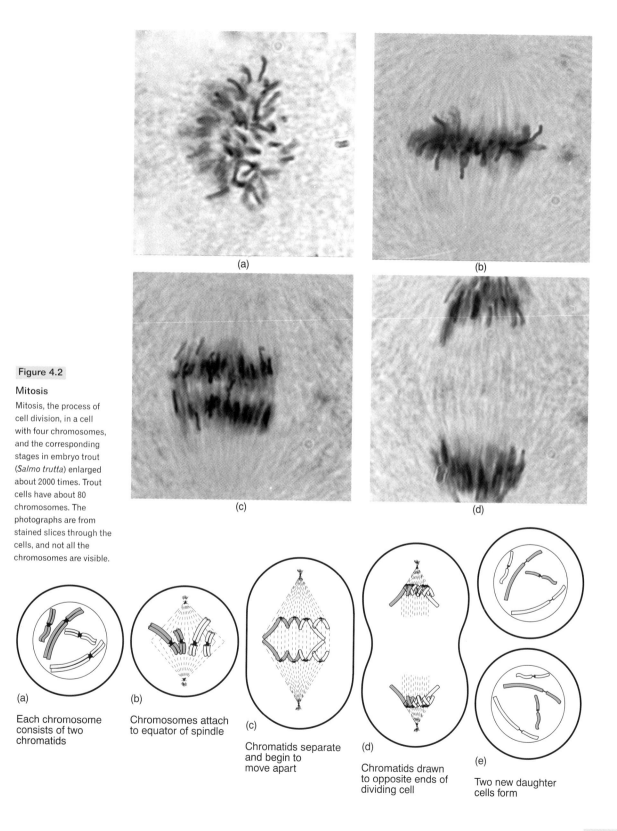

Figure 4.2

Mitosis

Mitosis, the process of cell division, in a cell with four chromosomes, and the corresponding stages in embryo trout (*Salmo trutta*) enlarged about 2000 times. Trout cells have about 80 chromosomes. The photographs are from stained slices through the cells, and not all the chromosomes are visible.

(a) Each chromosome consists of two chromatids

(b) Chromosomes attach to equator of spindle

(c) Chromatids separate and begin to move apart

(d) Chromatids drawn to opposite ends of dividing cell

(e) Two new daughter cells form

The first insight into the role of the nucleus came from study of its behaviour when the cell divides. Cell division is going on all the time in our bodies: growth, repair and replacement of tissues are all accomplished by increase in number of cells. In a resting (non-dividing) cell, the nucleus is a dense or dark globule with granular contents. When a cell enters the active, dividing phase, the contents of the nucleus become organized into a number of threads – the **chromosomes** ('coloured bodies', so named because of their affinity for the dyes used to make cell contents visible under the microscope). Each chromosome can be seen to consist of two separate threads (**chromatids**), held together at one point (Fig. 4.2(a)). The chromosomes then contract and thicken by coiling, the envelope bounding the nucleus disappears, and a spindle-shaped structure, formed of minute tubules, develops. The chromosomes attach to the central plane or equator of the spindle (Fig. 4.2(b)), the two chromatids composing each chromosome separate (Fig. 4.2(c)) and are drawn towards the poles of the spindle (Fig. 4.2(d)) as two sets of new, daughter chromosomes.

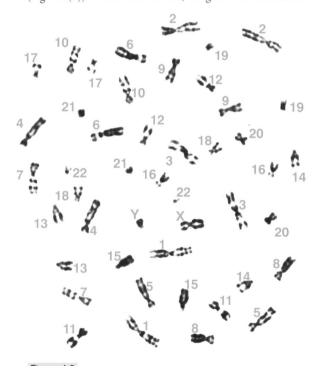

Figure 4.3

The chromosomes of a human cell

A human cell contains 22 pairs of chromosomes, matching in size and shape, and two chromosomes (X, Y) which do not match – 46 in all. The X and Y chromosomes are those responsible for sex determination.

Box 4.2

Sex determination

In all mammals, XY individuals are male and XX are female. All egg cells contain a single X chromosome, and males produce sperm which are 50% X-bearing and 50% Y-bearing. As a result, fertilized eggs are 50% XX and 50% XY, and the sex ratio is maintained.

Other groups of animals, such as birds and butterflies, have a different system in which males have a matching pair of sex chromosomes, and females have unlike sex chromosomes (butterflies) or only one sex chromosome (birds); in either case a 50:50 sex ratio results.

The spindle disappears, a new nuclear envelope appears around each of the new sets of chromosomes, and the whole cell divides into two (Fig. 4.2(e)). Finally, the chromosomes become less easily visible and replicate, each one producing two chromatids, during the resting stage.

This extraordinary sequence of events, called **mitosis**, happens every time a plant or animal cell divides. The number of chromosomes in dividing nuclei is constant for each species, and these numbers are even. Humans, for example, have 46 chromosomes in each cell (Fig. 4.3), while the small fruit fly *Drosophila*, the favourite experimental animal of geneticists, has six to twelve, depending on the species (see Fig. 6.6). It can also be seen that the chromosomes fall into pairs, the members of each pair being matched in size and shape. These regularities in number and behaviour of chromosomes suggested that they might be the carriers of the hereditary material. The matter was clinched by observations of chromosomes in the development of gametes (sperm and egg cells).

In the reproductive organs, the cells that are to produce gametes go through a cycle of division called **meiosis** ('lessening' or 'reduction'), which differs from the mitotic cycle of non-reproductive cells. Meiosis is shown in Fig. 4.4. When a sperm or egg cell enters the active phase of division, chromosomes become visible in the nucleus, as in mitosis, but each chromosome appears single, and is not split into two separate threads. Instead, the chromosomes associate in like-sized pairs, of the sort numbered in Fig. 4.3. Each chromosome now splits into two chromatids, as in early mitosis, so that each chromosome pair contains four threads, more or less tangled together. A spindle appears, the chromosome pairs attach to its equator and then separate, one of each pair being drawn towards each pole of the spindle.

Figure 4.4

Meiosis

Meiosis, the special cycle of cell
division that produces gametes
(egg and sperm cells). Further details
of meiosis are shown in Fig. 5.2.

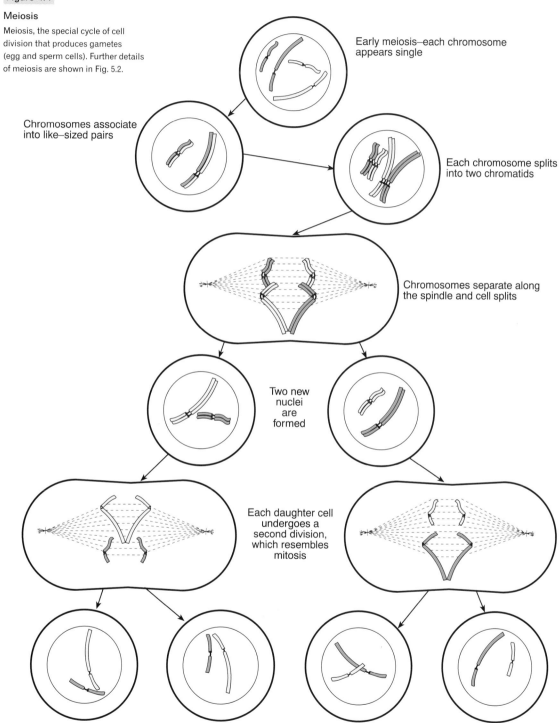

Early meiosis—each chromosome
appears single

Chromosomes associate
into like–sized pairs

Each chromosome splits
into two chromatids

Chromosomes separate along
the spindle and cell splits

Two new
nuclei
are
formed

Each daughter cell
undergoes a
second division,
which resembles
mitosis

Four daughter cells formed, each with half the number of chromosomes of the parent cell

The essential feature of meiosis is that in this first division the paired chromosomes (each containing two threads) move apart, whereas in mitosis it is the two threads of each chromosome that separate. After the first division two new nuclei are formed, each containing half the number of chromosomes of the parent nucleus – for example, four in *Drosophila melanogaster*, 23 in humans – and these chromosomes have already divided into two chromatids. The two new nuclei immediately undergo a second division which resembles mitosis in that individual chromosomes attach to the spindle, and the two chromatids of each are drawn apart. The end result of this meiotic cycle is four nuclei, each with half the number of chromosomes of the parent cell. In male organisms, each of these four nuclei develops into a sperm. In most females three of the nuclei degenerate and one becomes the unfertilized egg.

When fertilization occurs the egg cell is penetrated by a sperm and the nucleus of the sperm fuses with that of the egg. Because each of the two nuclei had half the number of chromosomes typical for the species, the resulting fertilized egg (**zygote**) has the full complement, half from each parent. This full complement of chromosomes is called the **diploid** number, and the half-complement found in gametes is the **haploid** number.

These complex nuclear manoeuvres are a perfect explanation of Mendel's theory, which was proposed before anything was known about chromosomes. Mendel guessed that each individual has a double dose of factors, that the factors segregate independently and at random in the gametes, and that they recombine in the next generation: this is an exact description of the behaviour of chromosomes. The fact that every organism has an even number of chromosomes in each nucleus and that these chromosomes fall into matched pairs (called **homologous** pairs) is due to the organism receiving one of each matched pair from the mother and one from the father.

The separation of these pairs in the first division of meiosis, one to each daughter nucleus, results in a random combination of maternal and paternal chromosomes, so that the four chromosomes in an egg cell of *Drosophila melanogaster*, for example, may contain one from the father and three from the mother, or two of each, or any other combination (exceptions to this rule are discussed in Section 5.4). The linkage between the inheritance of some genetic factors – the linkage groups mentioned earlier – is also explained: linked genes are to be found on the same chromosome.

The theory that the chromosomes carry the hereditary material became the basis of genetics for the first half of the twentieth century and was greatly refined by breeding experiments and observations of nuclear division. These refinements resulted in a concept of the chromosome as a thread with the genes arranged like beads along it. In the fly *Drosophila* the system of inheritance became so well understood, after many thousands of experiments, that 'maps' of each chromosome could be drawn with the relative positions of the genes for various characters marked. But there was still a great gap in genetic theory: what were these genes, and how did a gene control or direct the features of organisms? These questions were finally answered by work at the molecular level, in biochemistry rather than biology.

4.3   DNA and the genetic code

Shortly after the chromosome theory of heredity became established early in this century, the gross chemical composition of chromosomes was worked out. They were found to have two main components: **nucleic acids** and **proteins**. The nucleic acids were known to be relatively uniform compounds but proteins are extremely varied, and it was naturally thought that the hereditary material was the protein part of the chromosome. Although this idea turned out to be wrong, it is worth making a short digression on proteins.

Proteins are large, chain-like molecules built up of small subunits called **amino acids**, of which there are 20 different sorts. Natural protein molecules may contain any combination of these units, linked end to end (Fig. 4.5). The resulting protein chain, which may be of any length, is folded into a complicated three-dimensional pattern specific for each protein (Fig. 4.6). Because each link in the protein chain may be any one of the 20 different amino acids, the variety of possible proteins is inconceivably large – a protein molecule of 100 amino acids, a typical size for a natural protein, could have 20^{100} possible configurations, far more than the total number of atoms in the universe.

Proteins are the major component of living tissue. They include structural materials (such as keratin, which forms the hair of mammals and the feathers of birds, and collagen, which forms the framework of our bones and teeth) and transporters (e.g. haemoglobin, the red protein which carries oxygen in our blood). But the most important class of proteins is the **enzymes** – efficient biological catalysts that facilitate chemical reactions which would otherwise take place very slowly or not at all. Enzymes occur in

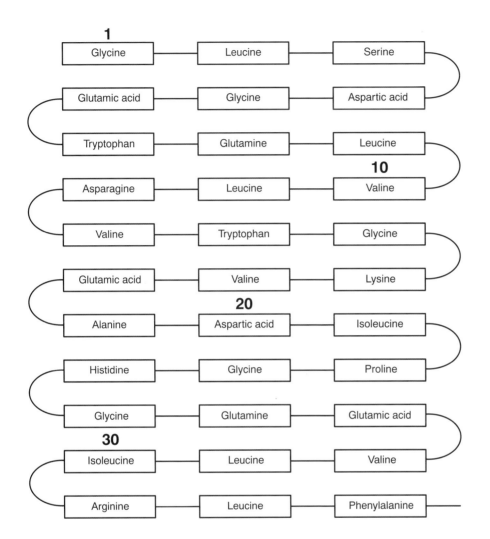

Figure 4.5

Structure of myoglobin
The first 33 amino acids of the human protein myoglobin.

organisms in enormous variety, each one specific for a certain reaction. Most of the life processes of organisms consist of chemical pathways, in which each step is a reaction mediated by a different enzyme. Some idea of the importance and variety of proteins comes from the estimate that each cell in our bodies contains at least 10 000 different types of protein.

Because of this enormous variety, proteins were obvious candidates for carrying genetic messages. But by the early 1950s sufficient experimental evidence had accumulated to show that the nucleic acids must be the hereditary material. Nucleic acids are of two types, **ribonucleic acid** (RNA) and **deoxyribonucleic acid** (DNA): RNA is mostly found outside the nucleus and DNA within the nucleus, in

the chromosomes. Both types are long molecules, with a chain-like backbone consisting of alternating phosphate and sugar subunits. (The sugar in RNA is ribose, in DNA deoxyribose.) Attached to each sugar subunit there is a base, the chemical term for a class of molecules including alkalis. DNA contains four different bases: adenine, cytosine, guanine and thymine.

Although the chemical composition of DNA was well known, there was no indication of how the substance could have a genetic function until 1953, when Francis Crick and James Watson proposed a model for the structure of the DNA molecule which showed how it could contain long, coded messages. This model – the double helix – with its biological implications ranks as the greatest contribution to

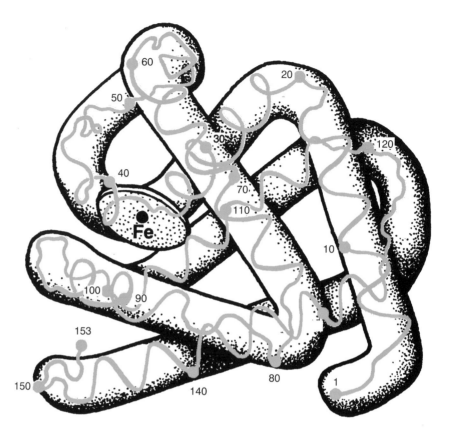

Figure 4.6

Structure of myoglobin

The three-dimensional structure of the protein myoglobin. The first level, the 'folded sausage', gives an idea of the form of the molecule and of the position of the oxygen-carrying haem group (the coin-like shape), containing an iron atom (Fe). The second level gives a more accurate representation of the chain of amino acids. The chain contains 153 amino acids (the first, last and every tenth one are numbered), and is partly spiral, partly angular or irregular.

biology since the work of Darwin and Mendel, something that is obvious enough from the fact that the acronym DNA and the image of the double helix are among the icons of late twentieth-century culture.

The Watson–Crick model of the DNA molecule can be thought of as a free-standing spiral staircase, in which the two strings or side-pieces are long chains of alternate sugars and phosphates, and each 'tread' is formed by two bases, linking the side chains (a simplified version is shown in Fig. 4.7). These treads are of just two types, because the four bases pair only in two ways – adenine always bonds to thymine and guanine to cytosine – but these base pairs may occur in any sequence, and may be oriented either way round (for example, in the adenine–thymine bond the adenine may be on the left or right side of the tread). This model has two most significant features.

- First, it offers a simple mechanism for replication (reproduction) of the molecule – progressive separation of the bonds between the bases 'unzips' and uncoils the double helix, allowing a new chain to be synthesized in sequence from sugar, phosphate and base molecules

(shown as building blocks ready for incorporation in Fig. 4.8) alongside each of the old side chains, which act as moulds or templates.

- Second, the model offers a method of carrying coded messages – the sequence of the bases. Along one side chain any one of the four bases may be attached to each sugar, and the sequence of the bases forms a coded message written in a four-letter alphabet, which we can symbolize by the initial letters of the bases: A (adenine); C (cytosine); G (guanine); T (thymine).

These implications of the model stimulated intense investigation, and within about a dozen years this code (the **genetic code**) had been cracked, and the way in which it is read and translated worked out. With a four-letter alphabet there are several ways of writing messages: if the letters are read individually, only four different statements will be possible; if they are read in pairs there are 16 possible statements (AC, AG, AT, AA, CT, CG, CC, CA, etc.); if read in threes 64, and so on. The genetic code is the same in almost all organisms, and uses non-overlapping triplets. For example, the base sequence TAGCATACT

Figure 4.7

Structure of DNA

Simplified version of the Watson–Crick model of DNA, showing the
sugar/phosphate side-pieces as spiral tubes and the two types of base-pair
crosspieces as interlocking planks. The proportions of the diagram (the
width of the spiral, ten base pairs to each complete turn) are correct, but
the chemical details are represented schematically.

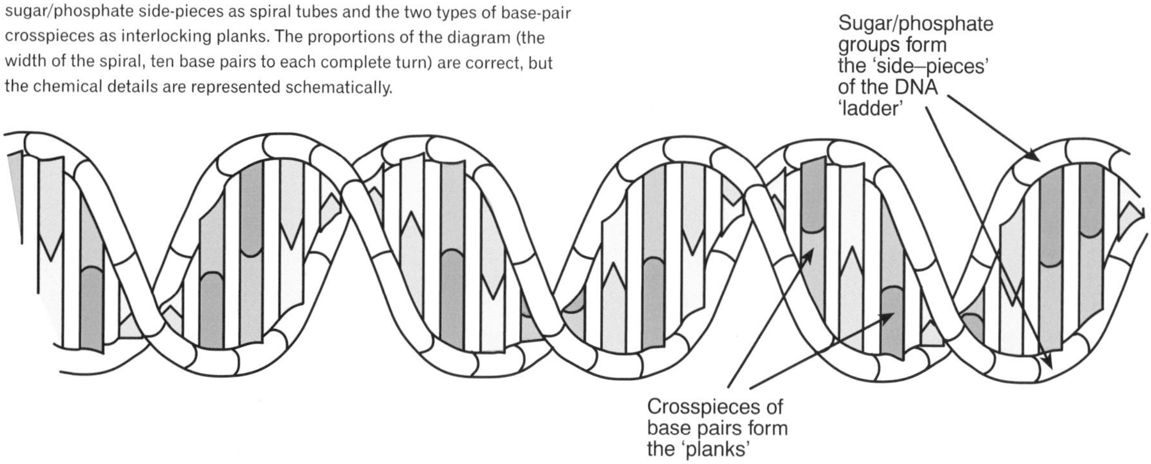

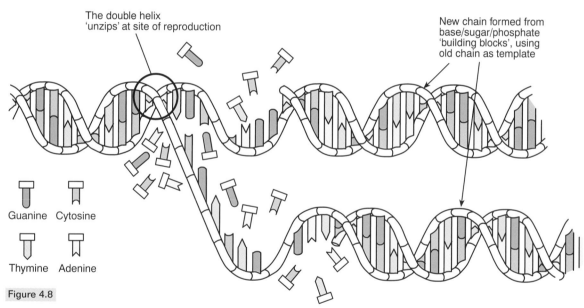

Sugar/phosphate
groups form
the 'side–pieces'
of the DNA
'ladder'

Crosspieces of
base pairs form
the 'planks'

The double helix
'unzips' at site of reproduction

New chain formed from
base/sugar/phosphate
'building blocks', using
old chain as template

Guanine   Cytosine

Thymine   Adenine

Figure 4.8

Reproduction or replication of the DNA molecule

Alongside each strand of the old molecule a new strand is forming from sugar,
phosphate and base molecules, shown as building blocks ready for incorporation.

would be read as the three words 'tag', 'cat' and 'act'. In a
code using triplets there are 64 possible statements. With
this limited language, it is obvious that these statements
must be very simple. In fact there are only two types of
statement: one specifies an amino acid, and one is a stop

sign. There are 20 different amino acids and a stop sign is
unambiguous, so the whole genetic language needs only
21 different statements. Most of the 64 possible messages
in the triplet code are therefore redundant, and several
have the same meaning. There are three different triplets

Exceptions to the genetic code

Intracellular organelles called mitochondria (see page 125) use a slightly different code in certain organisms (yeasts and mammals, for example) and two groups of single-celled organisms, the ciliates and mycoplasmas, use a code with one or two of the three normal 'stop' triplets assigned to an amino acid instead. But these exceptions do not alter the rule that the genetic code is the same everywhere, because they can be explained only as modifications of the universal code. An analogy would be the truth of the statement: 'all mammals have four legs' despite the fact that whales and dolphins have only two, because whales and dolphins can be understood only as mammals that have modified (lost or reduced) their hind legs.

Each genetic message must be contained in one strand of the double helix, because the other strand carries a complementary sequence of bases (T opposite A, G opposite C and vice versa) which will code quite a different message, presumably nonsensical. But when the molecule reproduces itself by the 'unzipping' procedure outlined above, a new meaningful message can be laid down on the mould or template formed by this nonsense strand. In the double helix, therefore, one strand carries the message and the other is necessary for replication of the molecule and the message.

Translation of the message is not direct – the amino acids are not assembled into proteins alongside the DNA. Instead there is a two-phase sequence: **transcription** (reading the message, in the nucleus) followed by **translation** (acting on the message by synthesizing protein, in the cytoplasm). Both processes involve RNA. RNA differs from DNA principally in having a single-stranded molecule, and in using the nucleic acid uracil (coded U) instead of thymine, so that the four letters in the RNA code are A, C, G and U. There are three varieties of RNA – messenger RNA (mRNA), ribosomal RNA (rRNA) and transfer RNA (tRNA). In the first phase of reading the message (transcription) an enzyme system binds to the DNA by recognizing promoter sequences near the beginning of the section coding a protein. The enzyme system transcribes this section of DNA, base by base, into a newly synthesized molecule of mRNA (Fig. 4.10) in which each base in the DNA is matched by its complement in the RNA (A matched by U, C by G, G by C, and T by A).

meaning stop, and the remaining 61 are distributed among the amino acids, with five of the amino acids coded by four different triplets and three by as many as six (Fig. 4.9).

The simplicity of this arrangement is surely the most stunning knowledge ever to have come out of biology. The only instructions in the genetic code are for assembling proteins. And the extraordinary variety of life seems ultimately to be due solely to differences in amino acid sequences. Genes for 'blue eyes' or 'wrinkled seed coat' do not exist: such features are the phenotypic consequences of a genotype that specifies nothing but proteins.

The way in which the DNA of a cell is read and translated is complicated, and will be described only in outline here.

First letter	Second letter				Third letter
	A	G	T	C	
A	AAA Phenylalanine AAG Phenylalanine AAT Leucine AAC Leucine	AGA Serine AGG Serine AGT Serine AGC Serine	ATA Tyrosine ATG Tyrosine ATT Stop ATC Stop	ACA Cysteine ACG Cysteine ACT Stop ACC Tryptophan	A G T C
G	GAA Leucine GAG Leucine GAT Leucine GAC Leucine	GGA Proline GGG Proline GGT Proline GGC Proline	GTA Histidine GTG Histidine GTT Glutamine GTC Glutamine	GCA Arginine GCG Arginine GCT Arginine GCC Arginine	A G T C
T	TAA Isoleucine TAG Isoleucine TAT Isoleucine TAC Methionine	TGA Threonine TGG Threonine TGT Threonine TGC Threonine	TTA Asparagine TTG Asparagine TTT Lysine TTC Lysine	TCA Serine TCG Serine TCT Arginine TCC Arginine	A G T C
C	CAA Valine CAG Valine CAT Valine CAC Valine	CGA Alanine CGG Alanine CGT Alanine CGC Alanine	CTA Aspartic acid CTG Aspartic acid CTT Glutamic acid CTC Glutamic acid	CCA Glycine CCG Glycine CCT Glycine CCC Glycine	A G T C

Figure 4.9

The genetic code
The 64 possible triplet combinations and their meanings. Most of the redundancy of the code is in the third letters of the triplets – for example GG– codes for proline, regardless of the third letter.

Figure 4.10

DNA makes RNA

An electron micrograph of DNA producing ribosomal RNA. The section of the DNA 'fibre' shown contains one gene for rRNA, out of which is growing 80–100 strands of RNA. These strands increase in length from one end of the gene to the other, the longest ones being fully transcribed, and the shortest just beginning transcription. The actual DNA and RNA molecules are not seen here, but the shape of these molecules is made visible by the proteins (enzymes and insulators) attached to them. The preparation is from the ovary of the clawed toad, *Xenopus*, enlarged about 28 500 times. Photograph courtesy of Dr Ulrich Scheer, German Cancer Research Institute, Heidelberg.

Because each amino acid in a protein is coded by a triplet of bases, we would expect a gene coding a particular protein to contain three times as many bases as the protein contains amino acids – a typical protein is about 150-200 amino acids long, implying typical genes about 550-700 bases long (allowing 100 bases for the promoter signals that precede the coding sequence). However, a typical gene is about ten times as long. A most surprising discovery, first made in the late 1970s, is that in most genes the section of DNA coding the protein is not continuous, but is interrupted by non-coding sequences (**introns**) that may be much longer than the coding sequences (the **exons**). The features of a typical gene are shown in Fig. 4.11.

The exons generally correspond to structural regions of the protein. Some introns are thousands of bases long. The newly synthesized molecule of messenger RNA reproduces the entire sequence of the gene, including the introns, and is then processed or 'matured' by snipping out the introns and splicing together the coding sequences. This splicing must be absolutely exact because the introns occur either between two triplets or inside one, and a mismatch of just one base would destroy the sense of the coding message. The processed molecule of mRNA, containing instructions for one protein, then passes out of the nucleus into the cytoplasm and associates with a **ribosome**.

Ribosomes can be visualized as sacs built of proteins and rRNA, and are the site of translation of the message. The threadlike mRNA molecule passes through the ribosome rather like recording tape through the head of a tape recorder, and as it passes through the triplet code is recognized by the third type of RNA, tRNA. Transfer RNAs constitute a family of molecules, each adapted to recognize a particular triplet of the code, and to add the appropriate amino acid to the protein chain, which emerges from the ribosome as it is synthesized, link by link (Fig. 4.12).

This is a much-simplified account of a complicated process, involving many different regulating proteins. Synthesis of a single protein molecule requires the presence of several hundred different proteins – acting as enzymes in transcription, mRNA processing, RNA and amino acid synthesis, rRNA and tRNA activity, and so on. And in every cell these proteins must have themselves been synthesized by the system, on the basis of information in the DNA, just as the transfer and ribosomal RNA must also be synthesized following information from the nucleus.

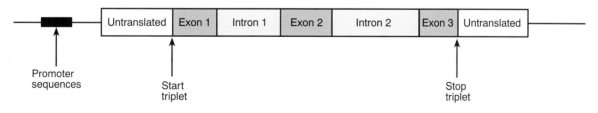

Figure 4.11

Structure of a typical gene

The DNA coding amino acids is interrupted by non-coding sequences called introns. The coding sequences, the exons, are typically about 150 bases long. Introns are much more varied in length and some are thousands of bases long. A typical protein is about 200 amino acids long, implying a gene of about 600 bases, but a typical gene is about ten times as long because of the introns and the untranslated regions at either end.

Figure 4.12

Translation of the genetic code to produce proteins

In nature, the ribosomes are positioned about 75 'spikes' apart on the mRNA molecule.

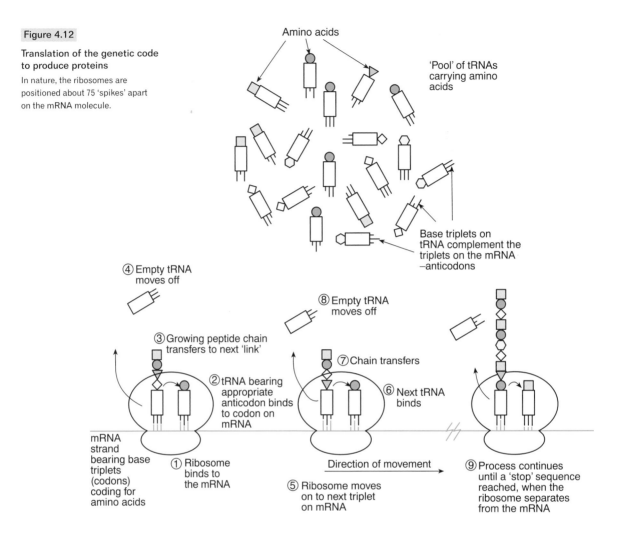

Amino acids

'Pool' of tRNAs carrying amino acids

Base triplets on tRNA complement the triplets on the mRNA –anticodons

④ Empty tRNA moves off

⑧ Empty tRNA moves off

③ Growing peptide chain transfers to next 'link'

⑦ Chain transfers

② tRNA bearing appropriate anticodon binds to codon on mRNA

⑥ Next tRNA binds

mRNA strand bearing base triplets (codons) coding for amino acids

① Ribosome binds to the mRNA

Direction of movement

⑤ Ribosome moves on to next triplet on mRNA

⑨ Process continues until a 'stop' sequence reached, when the ribosome separates from the mRNA

From this remarkable new knowledge of heredity at the molecular level we get a picture of the chromosomes as giant molecules of DNA, containing a series of recipes for proteins, with the individual recipes interrupted by introns, preceded by promoter sequences and separated by spacing sequences. This picture is still oversimplified, because in most organisms there are long sequences of repetitive DNA, with the same message repeated hundreds, thousands or millions of times, and also untranscribable regions of DNA, containing only nonsense.

Further, each cell in the body of an organism contains a full set of chromosomes, and therefore a full set of instructions for constructing the whole organism. Yet cells become specialized and differentiated; in animals some form muscle cells, others kidney cells, others nerve cells, and so on. So it is evident that these cells are acting only on a small part of the genetic message in their nuclei, and that the rest is not being read. Various mechanisms are known to prevent transcription of DNA: they can be thought of as switches and insulators, and most of them involve binding of protein with the DNA, to keep superfluous genes inactive. Production of these 'insulating' or 'repressor' proteins is itself the work of genes. There is probably a hierarchy of control genes of this type, some switching on and off a single gene, and some 'master genes' controlling batteries of genes.

The entire genetic message of an organism (the **genome**) was first worked out in 1976 for a virus that infects the human intestinal bacillus *Escherichia coli*. Viruses hardly merit the name 'organism', for they are capable of life only within the cells of their host, where they make use of the host's genetic machinery. Because of this parasitism at the molecular level, they can do without an elaborate protein-producing apparatus. The virus whose entire DNA was first decoded contains only nine genes, occupying 5386 code-letters (base pairs).

Two stretches of its DNA each contain two overlapping genes: two different messages are encoded in the same series of letters, using different reading frames. For example, the sequence CATTAGACTG can be read as 'CAT', 'TAG', 'ACT', 'G– –' if reading begins at the first letter, or as 'ATT', 'AGA', 'CTG' if reading begins at the second letter. This overlap could be seen as a way of economizing on DNA in minute organisms like viruses. During the 1980s and early 1990s larger viral genomes were decoded, containing up to about 200 000 base pairs.

In July 1995 the entire DNA sequence of the bacterium *Haemophilus influenzae*, 1.8 million base-pairs, was elucidated, followed three months later by the sequence of a second parasitic bacterium. In April 1996 the complete sequence (12 million base-pairs) of yeast was announced, and in August 1996 the first complete sequence of a free-living bacterium, *Methanococcus*, which has 1.7 million base-pairs and about 1700 genes, perhaps close to the minimum necessary for independent life. New complete genome sequences of micro-organisms now appear almost monthly.

The human genome project – directed for a time by James Watson of Watson–Crick fame – plans first to map and then to sequence the DNA of one human individual. The human genome contains about three thousand million (three billion) base pairs; by the end of 1997 the mapping work was virtually complete, and about sixty million base pairs had been sequenced. Similar projects are under way to sequence the DNA of other multicellular organisms – *Drosophila*, a nematode worm, and one or two plants.

The three billion base pairs of DNA in each of our own cells is sufficient to code at least a million proteins, allowing about 600 coding base pairs (200 triplets) per protein and another 2500 bases for introns and promoter signals.

Yet estimates from the human genome project and other sources are that about 80 000 genes are enough to specify a human being, and that only a tenth or less of our DNA consists of genes. What the rest of the DNA is doing is one of the interesting questions in evolution, and one we will return to in Chapter 10.

4.4   Summary

Modern molecular genetics has modified our previous ideas about heredity.

- We can contrast the Mendelian view of genes as entities that control single characters (like eye or flower colour) with the modern idea of genes as segments of DNA molecules that specify proteins, whose interactions may influence eye or flower colour (amongst other things).

- The designation of genes as dominant or recessive by Mendel and early geneticists must be modified. These categories referred to visible characters, which are not the direct product of the genes. In Mendel's experiments with peas, 'yellow' and 'green' must be due to differences in proteins – the 'AA' genotype producing one form, and the 'aa' another. In the 'Aa' genotype (yellow peas) the 'a' gene produces no visible effects, but at the molecular level the gene is presumably active. 'Recessive' is therefore a relative term: a recessive gene produces no visible effect, but may produce protein.

- The biological species concept – of interbreeding communities – can be supplemented (but not replaced) by a genetic species concept, of species as **gene pools** in which the genes reproduce asexually (when DNA replicates) and generate phenotypes (organisms) that can reproduce sexually, so producing new mixtures of genes.

- The mechanism of transcription and translation of the genetic code seems to be a one-way process: information passes from the DNA into the cell, but there is no known way in which the DNA can be modified by information from the cell. This has an important consequence, for it means that characters acquired during the life of an organism, such as the effects of use or disuse, cannot affect the DNA, and therefore cannot be inherited.

5 Genetics and variation

Any variation which is not inherited is unimportant for us.
Charles Darwin, *The Origin of Species*

With some understanding of the mechanism of heredity, we can look at variations within species in a new light, and answer some of the questions listed at the end of Chapter 3. Answering those questions will bring out other aspects of genetics.

First, the question of why organisms develop as near-perfect copies of their parents can be answered in terms of chromosome theory and the genetic code: the fertilized egg that develops into an individual contains a diploid set of chromosomes, half of the set received from each parent. These chromosomes contain coded information or a 'programme' which normally ensures that the individual is a copy of the parents.

5.1 Sterility of hybrids

Are there genetic mechanisms that allow some individuals to interbreed and prevent others from doing so? In other words, are there genetic reasons why different species do not hybridize?

A variety of such mechanisms is known, leading to failure of interspecific hybrids at various stages. If an egg is to be fertilized it must be penetrated by a sperm and the nuclei of these two cells must fuse. In many cases sperm of one species is unable to fertilize eggs of another. The mechanism here is usually chemical: the egg secretes a substance that repels, or does not secrete a substance that attracts, foreign sperm. Such chemical secretions are programmed by the DNA, and so are genetically controlled.

If fertilization is successful and a hybrid egg results, half its genetic message will specify a creature of one sort (the father) and half another (the mother) – it is not surprising that most such hybrids die early in development. But if the two sets of information are compatible and a hybrid, such as a mule, reaches maturity another genetic mechanism comes into play. This is the reduction division (meiosis) in

the reproductive cells, where the number of chromosomes is halved in gamete formation. Meiosis begins with the chromosomes associating in pairs, each pair containing one maternal and one paternal chromosome. This pairing is very exact, involving subunits within the chromosome, corresponding to sections of DNA. In hybrids, where the two sets of chromosomes differ, pairing may fail, or may be inaccurate or out of register. The result will be unequal distribution of the genetic material among the gametes – some will receive extra chromosomes or parts of chromosomes, others will lack them. If the parents have different numbers of chromosomes meiosis will always be irregular. This is one cause of sterility in mules, for the horse has 64 chromosomes (32 in gametes) and the donkey has 62 (31 in gametes), so a mule has 63 chromosomes and cannot pair them all off at meiosis.

5.2 Individuals are unique

Are the variations between individuals of a species inherited, or caused by variations in the environment and conditions of life? If they are inherited, what causes them?

This question – the extent to which individual variations are caused by heredity or the environment – is one of the oldest controversies in biology.

In order to determine the heritability of a feature, it is necessary to measure the total variability within the species and then, by observation or breeding experiments, to decide how much of that variation is caused by heredity and how much by different conditions. At one extreme are simple features such as the different colours of pea flowers and seeds used in Mendel's experiments, or the human ABO blood group system, which show no environmental effects and are determined entirely by heredity. At the other extreme are complex characters such as human intelligence, in which the two sources of variation are much harder to distinguish.

The extent to which individual variations are inherited has been most thoroughly studied in our own species. Identical twins, which develop from a single fertilized egg and so have the same genetic constitution, provide a good test. Studies of identical twins, especially the rare cases where twins are separated soon after birth and brought up under different circumstances, show that in most traits the greater part (60–100%) of the variation between individuals is inherited. Of course this is a general rule, not a law: if the wolf that found Romulus and Remus had suckled only one of the twins the great variation in their potential would have been determined entirely by the environment, not heredity. The continuing controversy over the genetic (inherited) and environmental (education) contributions to human intelligence shows how difficult it may be to make an objective estimate of the two components. Estimates of the genetic component of human IQ range from about 20% to 80%, and the argument continues about what is being measured and how bias is to be avoided. However, when the heritability of some uncontroversial feature, like bristles in *Drosophila*, is estimated at 80–90% no one troubles to argue.

The source of much of this heritable variation lies at the molecular level, in the base sequences of small sections of DNA. Mendel assumed that individuals have a double dose of each heritable factor, and that these factors might be both the same, as in his true-breeding yellow peas (AA), or different, as in yellow peas with mixed green and yellow offspring (Aa). In modern terms the segment of DNA controlling the factor in question should either be the same in the chromosomes received from each parent or different. The idea that there are only two alternative forms of a gene (**alleles**) is an over-simplification – many, perhaps most, factors have three or more variants within a species. Mendelian breeding experiments may show that they are all alternative forms of the same section of DNA: this will be so if no individual is found to possess more than two of the possible variants. Such sets of variants – alternative forms of the same part of the chromosome – are called **multiple alleles**.

5.3 Multiple alleles

Where a character, such as height or intelligence in human beings, is controlled by many genes, each of which may have several allelic forms, the variety of possible genotypes

is very large. A simple example is the human ABO blood group system, mentioned in Chapter 3 as an instance of polymorphism. The ABO system is controlled by four alleles, symbolized A_1, A_2, B and O: A and B genes are dominant to O (recessive), A_1 is dominant to A_2 and the two A genes are neither dominant nor recessive to B (both alleles are expressed in the phenotype of the AB genotype). The ten possible genetic constitutions of individuals with these four alleles are:

$A_1A_1, A_1A_2, A_1B, A_1O, A_2A_2, A_2B, A_2O, BB, BO$ and OO.

If there were five alleles, the number of possible genotypes would be 15, for six alleles 21, and so on.

Multiple alleles introduce another very important idea – that one individual is never an adequate sample of the genes available in a species (Fig. 5.1). No individual can have more than two alleles of any gene (one in each of a pair of chromosomes), but the species may contain many sets of multiple alleles.

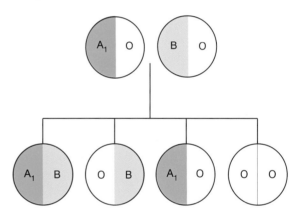

Figure 5.1

Multiple alleles: the human ABO system
The parents have three of the four alleles for the gene between them. Thus there are four possible blood groups for the children to inherit.

The main source of the genetic uniqueness of an individual is reshuffling of parental genes. One cause of this shuffling is random segregation or sorting of chromosomes in meiosis: for example an organism with 20 chromosomes, ten from each parent, will produce gametes containing ten chromosomes with every possible assortment of paternal and maternal chromosomes. But in meiosis an additional source of variation (omitted, for the sake of simplicity, in the account in Section 4.2) is the phenomenon called **crossing-over**.

Figure 5.2

Crossing-over

The simplest type of crossing-over in a pair of homologous chromosomes. Two or more crossovers often occur so that two or more sections of chromosome are exchanged. If no crossovers occur the pair of chromosomes will not remain together and meiosis will be abnormal.

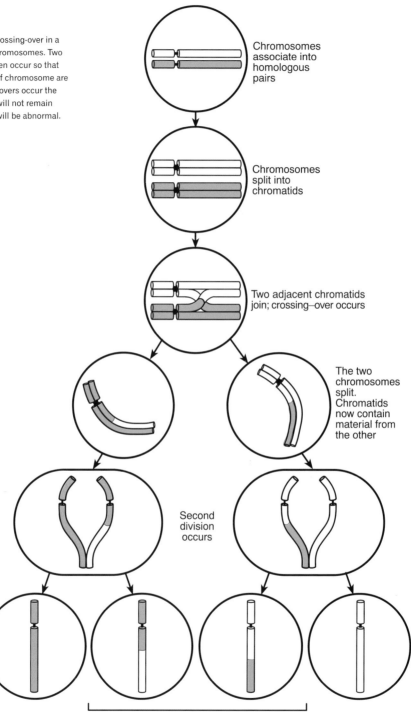

Chromosomes associate into homologous pairs

Chromosomes split into chromatids

Two adjacent chromatids join; crossing–over occurs

The two chromosomes split. Chromatids now contain material from the other

Second division occurs

New chromosomes formed by cross–over

5.4 Crossing-over

In the first phase of meiosis pairs of homologous (one maternal and one paternal) chromosomes associate and then each chromosome splits into two chromatids, so that there are four associated threads (see Fig. 4.4). A most surprising finding, which has been confirmed by direct observation and by breeding experiments, is that these threads are not simply tangled like four lengths of string, where the four separate pieces can be recovered by untangling. The association involves actual breakage and rejoining of the threads, with mutual exchange (crossover) of sections of chromosome. Continuing the string analogy, it is as if one tangled up two pieces of green string and two of white and found on separating them that one of the green pieces is the same length as before but includes one or more white sections, while the missing green sections are incorporated in the appropriate places in one of the white pieces. If such a thing happened with string we would take it to be a miracle or, more probably, a conjuring trick. But with chromosomes in meiosis it is a regular (though not predictable) occurrence.

DNA in early meiosis behaves as if it were easily broken and as if the loose ends were 'sticky', so that a broken piece of DNA will unite with any other available loose end. When two DNA molecules are lying very closely side by side, part of one side chain in each double helix may break free from its original partner and associate with the other molecule. Repair of the breaks will lead to exchange of sections of DNA, of varying size, between chromosome pairs (Fig. 5.2).

Thus, because of crossing-over, after meiosis the gametes receive an assortment of mixed chromosomes with varying combinations of maternal and paternal segments. These mixing processes generate an almost infinite variety of gametes and, because every individual is produced by fusion of two gametes (each resulting from shuffling of a different set of genes), it would be surprising indeed if the individual genotype were not unique.

But this shuffling of pre-existing DNA is not the end of the matter. In Section 3.1, human fingerprints and 'chemical fingerprints' were mentioned as examples of individual variation. During the last few years 'DNA fingerprinting' has been applied in forensic science, in legal disputes over paternity and investigations of heritable disease within our own species and in population biology of other species. DNA fingerprinting depends on crossing-over within **minisatellite** DNA – sections of the chromosome consisting of many repeated copies of a short sequence (10–50 bases) of non-coding DNA. These multiple repeats seem to act as 'hot spots' in crossing-over and the crossing-over is often unequal so that, if each chromosome of a maternal and paternal pair had 30 repeats in one minisatellite, after crossing-over one of the two descendant chromosomes might have 20 and the other 40. Minisatellite DNAs are thus highly variable. In fact it is unlikely that both chromosomes in a pair would have exactly the same number of repeats in each minisatellite, and imperfect matching between the two variants will be the main cause of unequal crossing-over. Each individual therefore has a unique pattern of repeats in minisatellite DNA, and a single hair or sperm may be enough to identify that individual.

DNA fingerprinting has also been used to study patterns of descent. For example, among 34 offspring produced in 11 broods by seven different female house sparrows, two chicks each had a different father from the rest of their brood, and one brood was produced by mother/son mating. In human terms, adultery and incest are evidently quite common in sparrows, but the general point here is the precision with which a human or a sparrow can be individualized by a scrap of DNA. This illustrates the generation of variability during meiosis and brings in mutation, which is discussed in the next chapter.

6 Mutation

Selective breeding in domesticated animals and plants (Section 3.4) shows that the potential variability of a single wild species is enormous, and raises the question of the origin of heritable variation. New combinations of genes present in wild populations and the occasional appearance (and selection by breeders) of individuals showing rare recessive traits account for much of this variation. However, there is another important source – mutation, or the appearance of novelties. In everyday usage, as in science fiction, the words 'mutation' and 'mutant' imply striking and obvious changes, and this was the original meaning of the terms in genetics. But in the modern molecular theory of genetics, mutation means the occurrence of any change in the genotype, and many known mutations have no detectable effects: they are not manifested in the phenotype.

Mutations are of two main sorts:

- Changes in the coding of small sections of DNA, within individual genes – **point mutations**.
- Large-scale changes in the chromosomes – **chromosome mutations**.

6.1   Point mutations

With knowledge of the code in which the genetic message is 'written', theoretical understanding of point mutations is fairly simple. Imagine a short section of DNA within a gene specifying an enzyme, in which the code (symbolized by the initials of the four bases; see Fig. 4.9) reads:

CAT TAG GAT ACT

If the first symbol in this series happens to be the first of a triplet these bases will form a sequence of four triplets. After transcription and translation they will specify the amino acids valine, isoleucine and leucine and a stop sign.

Suppose that by some accident in replication the first base pair in this section of DNA were to be eliminated.

The message now reads:

ATT AGG ATA CT

The first triplet has become ATT, a stop sign, so the remainder of the code will not be translated and the enzyme molecule will lack its last three amino acids. This type of mutation is known as **deletion**.

Alternatively, suppose that an extra base, say G, were inserted immediately before the first C in the original sequence. It now reads:

GCA TTA GGA TAC T

This specifies arginine, asparagine, proline and methionine, and there is no terminal stop sign, so the following DNA will also be translated and added to the enzyme molecule. This is an **insertion** mutation. Deletion (of one or more bases) and insertion (of one or more bases) mutations will clearly have profound effects. But notice that if an insertion or deletion involved three base pairs, not just one, the effect would be small, adding or deleting a single amino acid in the protein chain. Deletions or insertions of base pairs in numbers other than three or multiples of three are called **frameshift** mutations because they alter the reading frame of the DNA.

A more common type of mutation is a **substitution**, in which one base pair is replaced by a different one. Suppose the first of the two Gs in the middle of the sequence is replaced by T, so that the triplet TAG is changed to TAT. Reference to the code shows that both triplets specify isoleucine, so such a mutation would have no effect on the protein (mutations of this kind are called **silent** or **synonymous**). Most substitutions within coding sections of DNA will alter one amino acid into another at a certain point in the protein chain, and may or may not affect the functioning of that protein; substitutions that alter amino acids are **missense** mutations. The most drastic

Box 6.1

Mutations in human beta haemoglobin

The normal DNA sequence, including that coding the last few amino acids, is shown in Fig. 6.1. Many mutants of this sequence have been discovered: each is named for the place it was first detected.

The 'McKees Rocks' mutation (Fig. 6.2) was first seen in a family in the USA in 1976. The ATA triplet that codes tyrosine at position 145 has been changed to a stop triplet (ATT) by a substitution mutation, and the haemoglobin chain is cut short at position 144 so it lacks the last two amino acids. A more serious change to 'stop' has been seen in the the DNA of a 28-year-old man. A substitution mutation in the 17th triplet cut the protein so short that it was unrecognizable. This man had no beta haemoglobin at all.

In the 'Tak' mutation (Fig. 6.3) a base has been deleted from the GTG triplet at the end of the chain. This alters the reading frame from that point onwards, producing threonine at position 146 (rather than histidine). The new stop (ATT)

sequence terminates the chain ten amino acids later than normal.

The third example is the 'Saverne' mutation (Fig. 6.4), first seen in a 34-year-old French woman in 1983. The second base in the GTG triplet at position 143 has been deleted, altering the reading frame from there on. The last four amino acids (from position 143 onwards), have been altered and an extra ten amino acids added to the end of the protein, until a stop sequence is reached.

Other deletions noted in the beta chain involve removal of triplets, giving haemoglobin chains lacking from one to five amino acids.

The incidence of these mutations is very variable: some are widely distributed but some have been seen in only one family. It is important to note that each of these mutations is found in both children and adults – none inhibits development and most produce no detectable physical symptoms.

CGG GAC CGG GTG TTC ATA GTG ATTCGAGCGAAAGAACGACAGGTTAAAGATAATTTCCAAGGAAA........
140　141　142　143　144　145　146
Ala　Leu　Ala　His　Lys　Tyr　His

Figure 6.1

Normal human beta haemoglobin

The normal DNA sequence for the last few amino acids of beta haemoglobin.
The stop triplet is shown in coloured type in each case.

CGG GAC CGG GTG TTC ATA GTG ATTCGAGCGAAAGAACGACAGGTTAAAGATAATTTCCAAGGAAA........
CGG GAC CGG GTG TTC AT☐ GTG ATTCGAGCGAAAGAACGACAGGTTAAAGATAATTTCCAAGGAAA........
Ala　Leu　Ala　His　Lys

Figure 6.2

A substitution mutation in human beta haemoglobin

The 'McKees Rocks' mutation (boxed). The ATA triplet at position 145 changes into a stop signal, cutting the protein chain short.

CGG GAC CGG GTG TTC ATA GTGA TTC GAG CGA AAG AAC GAC AGG TTA AAG ATA ATTTCCAAGGAAA........
CGG GAC CGG GTG TTC ATA☐TGA TTC GAG CGA AAG AAC GAC AGG TTA AAG ATA ATTTCCAAGGAAA........
Ala　Leu　Ala　His　Lys　Tyr　Thr　Lys　Leu　Ala　Phe　Leu　Leu　Ser　Asn　Phe　Tyr

Figure 6.3

A frameshift mutation in human beta haemoglobin

In the 'Tak' mutation deletion (boxed) from the triplet at position 146 causes an extra ten amino acids to be added to the protein chain.

CGG GAC CGG GTGT TCA TAG TGA TTC GAG CGA AAG AAC GAC AGG TTA AAG ATA ATTTCCAAGGAAA........
CGG GAC CGG G☐GT TCA TAG TGA TTC GAG CGA AAG AAC GAC AGG TTA AAG ATA ATTTCCAAGGAAA........
Ala　Leu　Ala　Pro　Ser　Ile　Thr　Lys　Leu　Ala　Phe　Leu　Leu　Ser　Asn　Phe　Tyr

Figure 6.4

A frameshift mutation in human beta haemoglobin

The 'Saverne' mutation (boxed): deletion of the second base in the triplet at position 143 has altered the amino acids added to the protein chain from then on, and the chain is ten amino acids longer than normal.

substitutions change a triplet specifying an amino acid into a stop sign (**nonsense** mutations), or change a stop sign into an amino acid triplet – the first of these would cut the protein short, and the second would add new amino acids to the end of the chain.

All of these types of mutation are found in nature. A simple example is seen in the human ABO blood groups (Sections 3.3 and 5.3). The ABO gene codes a 350 amino-acid enzyme involved in synthesis of A or B antigens. The A and B alleles differ by 5–10 point mutations, but because several of those are silent the proteins produced by A and B alleles differ by only four amino acids. In the O allele a single base has been deleted in the 87th triplet, a frameshift mutation that changes the reading frame. The O protein differs from A and B in amino acids 87–116, and then has a 'stop' triplet, so that the protein is only one-third as long as the A or B proteins. This truncated O protein is non-functional and in OO individuals no antigen is produced.

Other well-known examples occur in human haemoglobin. The human haemoglobin molecule consists of two pairs of protein chains linked together. The two types of chain, alpha and beta, are each similar in form to the myoglobin molecule shown in Fig. 4.6. The two are specified by genes on different chromosomes. The normal alpha chain contains 141 amino acids, the beta chain 146. By 1994 more than 550 different mutations in human haemoglobin had been recorded, including 182 substitutions in the alpha chain, 316 substitutions in the beta chain, three deletions in the alpha chain, 14 deletions in the beta, three insertions in the alpha chain and four in the beta. All except 17 of the 500 substitution mutations can be accounted for by a change in a single base pair in a DNA triplet; two of the exceptional 17 require changes at two positions in a triplet, and the other 15 require one change in two different triplets. Four alpha chain and one beta chain substitutions are in the stop triplet and add up to 31 extra amino acids to the chain (an example is the 'Tak' mutation; Fig. 6.3). Two beta chain substitutions produce stop triplets: one in the last triplet but one ('McKees Rocks'; Fig. 6.2); the other occurs in the seventeenth triplet and cuts the protein short so that it is unrecognizable. One alpha chain and two beta chain deletions are frameshift mutations, removing a single base-pair (about ten bases before the normal stop triplet) and altering the reading frame so that the usual last three amino acids are changed, the stop sign is missed, and extra amino acids are added to the chain (the 'Saverne' mutation, Fig. 6.4). Mutations in beta haemoglobin are discussed in more detail in Box 6.1.

The mutations in human haemoglobin are not randomly distributed through the protein chains: some parts of the chain are prone to mutation and others are not. In the beta chain, the 330 known substitution and deletion mutations affect all but eight of the 146 amino acids in the chain.

Amongst these amino acids that seem to be immune to change, one might guess that some have merely been lucky so far and will be found to be subject to change (between 1985 and 1989 the number of 'immune' amino acids in the beta chain dropped from 21 to 14, and between 1989 and 1994 it dropped to eight) whereas others are so vital to the function of the molecule that mutations affecting them are lethal, impairing haemoglobin function so seriously that they cause early death. These vital parts would include the groups that give the molecule its three-dimensional shape, those that join the alpha and beta chains together, and the region surrounding the haem group, which carries oxygen. Mutations are not less likely in these regions, but bearers of such mutations do not survive to manifest them. Many lethal mutations are known, and where the precise genetic cause can be elucidated it is usually a point mutation.

About one person in 2000 carries a mutant haemoglobin gene: a few of these have newly arisen within that person, the rest have been inherited. The basic cause of these mutations is error in replication of DNA, the newly formed DNA molecule being an imperfect copy of the parent molecule. It is important to find out how frequent such mistakes are – what the **mutation** rate is.

The rate at which mutations appear is greatly increased by treatment with X-rays and other radiation, such as gamma rays and ultraviolet light. The rate is also increased by various chemicals, including mustard gas and hydrogen peroxide. In nature, the most common cause of mutation is probably the background radiation to which all organisms are subject, from cosmic rays, natural radioactivity and so on. The natural, or spontaneous, mutation rate has been calculated for a variety of different mutations in animals, plants and micro-organisms. There are two ways of expressing this rate:

- as mutations per gene – mutations in the section of DNA that specifies a particular protein, or controls a visible character;
- as mutations per base pair in the DNA.

The spontaneous mutation rate per gene, calculated for mutations with a visible effect, not for 'hidden' ones like those haemoglobin mutations which produce no physical symptoms, is about one in 100 000 per generation, for a wide variety of organisms. Each gene contains many coding base pairs: for example, the genes coding haemoglobin chains (with about 150 amino acids) contain about 450 coding base pairs because each amino acid is coded by three base pairs. The mutation rate per base pair is therefore much lower than the gene rate, and a base pair rate of about one in 100 million per generation is generally accepted. If this rate is constant, as it seems to be, 'long' genes specifying long protein chains (some contain 600 amino acids) will have a higher mutation rate than 'short' genes.

Because a mutation may occur whenever DNA replicates, and since cell division, which involves DNA replication, is going on all the time, we accumulate mutations throughout life. Mutations are rare events, but our bodies

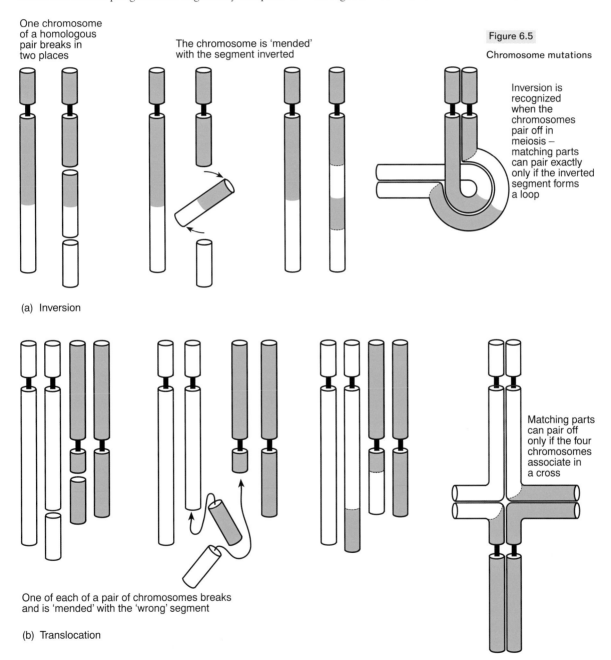

One chromosome of a homologous pair breaks in two places

The chromosome is 'mended' with the segment inverted

Figure 6.5

Chromosome mutations

Inversion is recognized when the chromosomes pair off in meiosis – matching parts can pair exactly only if the inverted segment forms a loop

(a) Inversion

One of each of a pair of chromosomes breaks and is 'mended' with the 'wrong' segment

Matching parts can pair off only if the four chromosomes associate in a cross

(b) Translocation

contain so many dividing cells that every human probably carries hundreds of base pair mutations that have arisen during his or her life.

6.2   Chromosome mutations

The point mutations just discussed usually involve changes in single base pairs in DNA – they are very small-scale phenomena. Chromosome mutations involve changes in large pieces of DNA – segments of chromosomes or even whole chromosomes – and, while point mutations imply actual change in the genetic material, chromosome mutations usually rearrange existing material. Chromosome mutations can be provoked by the same agents as point mutations – radiation and chemicals – but also occur spontaneously. Point mutations occur when DNA reproduces itself, and chromosome mutations occur when the nucleus reproduces itself in cell division, especially in meiosis.

The phenomenon of crossing-over, a peculiarity of the meiotic cycle which results in exchange of matching sections between homologous chromosomes was described in Section 5.4; DNA behaves as if it were easily broken, and as if the loose ends were sticky. Chromosome mutations are mostly due to errors or mispairings in crossing-over. They are of seven main types:

- **inversions**, where a section of a chromosome maintains its position, but is turned round end to end (Fig. 6.5(a));
- **translocations**, where sections of chromosome are exchanged between non-homologous chromosomes, not between homologues, as in normal crossing-over (Fig. 6.5(b));
- **deletions** and **duplications**, which arise when crossing-over is unequal, one chromosome of the pair receiving more than it loses, so that a part of it is duplicated, while a part is deleted from its partner;
- **fusions,** in which two whole or almost whole chromosomes join end to end, reducing the number of separate chromosomes by one;
- **fragmentations**, where a chromosome splits into two, increasing the number by one;
- **unequal divisions**, in which after a cycle of cell division one of the two daughter nuclei receives one or more extra chromosomes and the other lacks one or more.

Examples of all these types of chromosome mutation are known. Inversions and translocations can often be recognized under the microscope by the pattern of banding of the chromosomes, or by unusual pairing patterns in cells undergoing meiosis (Fig. 6.5). In our own species, to take one example, 3% of the population of Edinburgh carry an inversion of part of chromosome 1. The standard human set of chromosomes differs from that of chimpanzees by eight inversions, and from gorilla by eleven inversions and a translocation. In meiosis, inversions form loops and translocations form crosses (Fig. 6.5), complications that often lead to gametes lacking segments of chromosomes and so to reduced fertility. Inversions and translocations are therefore genetic changes that may prevent or inhibit hybridization between species. However, in many animal species inversions are so widespread that they are clearly not harmful – in *Drosophila* the frequency of different inversion patterns varies with the seasons, in an annual cycle that seems to be maintained by natural selection. The explanation for this may be that the inverted segments act as 'supergenes' (the looping in meiosis means that products of crossing-over within the inversion do not survive), so that a block of genes can be kept together.

Deletions and duplications can range in scale from a few base pairs (like the deletions from human beta haemoglobin: Box 6.1) up to visible chunks of chromosome. Large-scale deletions of chromosome segments are often lethal; duplications, on the other hand, may be an important source of evolutionary innovations (see Sections 10.2 and 13.3). On the small scale, DNA fingerprinting (mentioned in the previous chapter as an illustration of the generation of variability by unequal crossing-over – chromosome mutation by mutual deletion and duplication in the minisatellite DNA of a pair of chromosomes) has shown that in one of the minisatellites with a short repeat length (nine base pairs) the detectable mutation rate is about 5% per gamete. Since deletions or insertions of less than about five repeats (5 x 9 = 45 base pairs) would not be detectable, the mutation rate may be 10% per gamete, the highest mutation rate yet estimated.

Fusions and unequal divisions seem to have played a conspicuous part in evolution, otherwise it is hard to see why different species have different numbers of chromosomes. For example, the difference between a donkey, with a diploid set of 62 chromosomes, and a horse with 64, or between a human (46) and a gorilla or chimpanzee (48), is most probably due to some past event of fusion in the human line and fragmentation in the horse line. A simple example of chromosome fusion in species of *Drosophila* is shown in Fig. 6.6. A more extreme example among mammals is seen in muntjacs, small south-east Asian deer. The Chinese muntjac, *Muntiacus reevesi*, has 46 chromosomes and the Indian muntjac, *Muntiacus muntjak*, has only six (in females) or seven (in males). Despite this huge disparity (evidently due to fusions in the Indian muntjac lineage), the two muntjac species can produce viable hybrids, which receive three or four chromosomes from one parent and 23 from the other. As might be expected, the hybrids are sterile, but the example illustrates that chromosome fusion may make little difference to the genotype because it involves no gain or loss of material.

Unequal division, producing cells lacking or with extra chromosomes, is much more drastic. Lack of a chromosome is usually lethal (as might be expected) but extra chromosomes also produce serious disturbances. One or two children in every thousand are afflicted by Down's syndrome, a malfunction in which the face has a characteristic cast, mental development is retarded, males are sterile and females very infertile. This is caused by an extra chromosome matching one of pair 21 in the standard set (Fig. 4.3), so that a victim of Down's syndrome has 47 chromosomes, with three number 21s.

6.3   Polyploidy

The most extreme example of unequal cell division is doubling of the chromosome number, which will occur if division of the chromosomes is not followed by division of the nucleus. The normal number of chromosomes for a species is the **diploid** number, while the half-set found in gametes is **haploid**. Doubling of the diploid set produces cells with a **tetraploid** complement (Greek *tetra* = four), and a cross between a diploid organism (with haploid gametes) and a tetraploid (with diploid gametes) would produce a **triploid**, with three times the haploid number. Organisms with chromosome numbers which are some multiple of a basic set are called polyploids. **Polyploidy** is uncommon in animals, but it has been of great importance in plant evolution: almost half the known species of flowering plants are polyploids.

Suppose that, in a plant, abnormal cell division in a developing bud or flower produces a cell with double the normal number of chromosomes. This cell may continue to divide, and all its descendants will also have a double set of chromosomes. In this way, a tetraploid flower or branch may develop and produce pollen and egg cells that are diploid – not haploid, as is usual. Self-fertilization of these flowers would produce a batch of tetraploid seeds. Such events are improbable in normal plants, because meiosis (the reduction division in the male and female parts of the flower) will be disturbed in the tetraploid cells since the chromosomes will tend to associate in fours rather than pairs, leading to irregular chromosome numbers in the gametes and to reduced fertility. But suppose the plant were a hybrid between two different species, one with a diploid number of (say) eight, the other with 12. Such a hybrid would have ten chromosomes, four from one parent

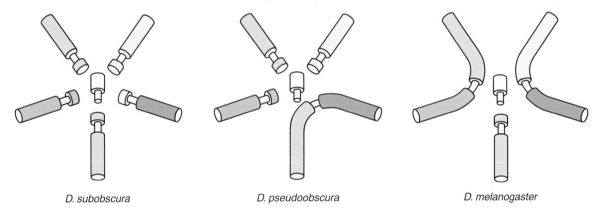

| *D. subobscura* | *D. pseudoobscura* | *D. melanogaster* |

Figure 6.6

Chromosome fusions in *Drosophila*

Chromosome fusions of haploid cells (gametes) of the fruit fly *Drosophila*. *Drosophila subobscura* has one dot-like and five rod-like chromosomes. The pattern in *Drosophila pseudoobscura* can be derived by fusion between two chromosomes, and that in *Drosophila melanogaster* by two such fusions.

and six from the other, and would be completely sterile because the chromosomes cannot pair off at meiosis. But if a tetraploid flower arose in this hybrid plant it would be fertile, because the 20 chromosomes would fall into ten homologous (matching) pairs. These flowers, if self-fertilized, will produce seeds combining the genotypes of the two parent species. What is more, the plants growing from these seeds will be interfertile, forming the basis of a reproductive community – but any hybrids formed with either of the parent species would be triploid and infertile. By this sort of accident an entirely new species of plant can arise more or less instantaneously.

Figure 6.7

Polyploidy in wheat

Bread wheat, *Triticum aestivum*, has evolved by successive chromosome doubling in hybrids between different species.

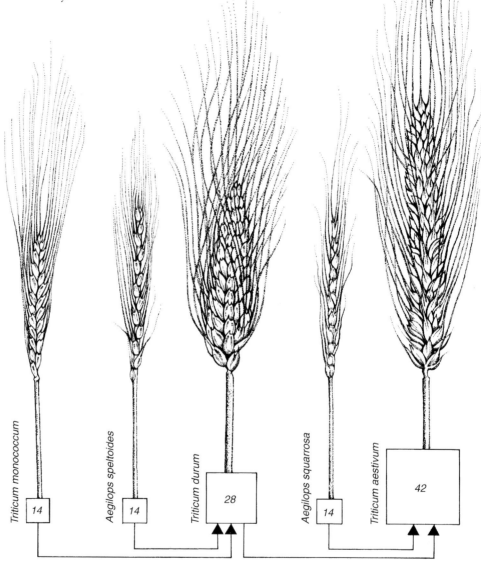

Polyploids can be produced quite readily in the laboratory by treating plants with colchicine, a substance extracted from *Colchicum* (the autumn crocus) which interferes with cell division so that the chromosomes divide but the nucleus does not. Polyploid plants differ from diploids in having larger cells (because the nucleus is larger), thicker, fleshier leaves and larger flowers and fruit. Fleshier leaves and larger fruit are desirable features in plants cultivated for food, so it is not surprising that many of our staple food plants are polyploids.

Perhaps the most important example is wheat (Fig. 6.7). The wheat first cultivated by early farmers was a diploid species, *Triticum monococcum* or einkorn wheat, with 14 chromosomes. It is still found wild in the Near East. *Triticum durum*, the hard or 'macaroni' wheat now cultivated for pasta, is a tetraploid species, with 28 chromosomes. It arose by chromosome doubling in a hybrid between einkorn wheat and a grass, *Aegilops speltoides*, also with 14 chromosomes, which grows as a weed in fields in the Near East. This tetraploid wheat may have been found wild, or may have first appeared and been selected, in some early wheat field. The wheat now grown for bread, *Triticum aestivum*, is a hexaploid (Greek *hexa* = six) species with 42 chromosomes, which arose by a second episode of chromosome doubling in a hybrid between durum wheat and another grass, *Aegilops squarrosa*, with 14 chromosomes. Durum and bread wheats have been 'recreated' in the laboratory by inducing chromosome doubling with colchicine in hybrids between the three wild species.

Many other cultivated plants are polyploids, and the production of new varieties with increased yields by inducing chromosome doubling is an important branch of agricultural research.

In plants, new species can arise instantaneously by chromosome doubling and new species can be 'created' in the laboratory, but this process alone will not explain evolution, for polyploid species are rare in animals. This is not because the nuclear accidents causing chromosome doubling do not happen in animals: examination of the chromosomes of 227 human embryos that died before birth showed that two were tetraploids, and in many animals, especially insects, certain organs of the body are always formed by polyploid cells, in which nuclear division has stopped but chromosome division continues. The main reason for the rarity of polyploid species in animals is that most animals are not hermaphrodites capable of self-fertilization – and self-fertilization, common in plants, is necessary for successful species formation by polyploidy. The only animal groups in which polyploid species are frequent are hermaphrodite worms, and some crustaceans, insects and a few other animals that reproduce asexually. Yet there are polyploid species and groups of species in some sexual animals – certain frogs and fishes, for example. Occasional chromosome doubling may explain some of the millionfold variation in the DNA content of the cell observed in different organisms (Fig. 6.8). The rest is explained by the accumulation of introns and other non-coding DNA (page 21) and 'junk' DNA (page 76).

Figure 6.8

The amount of DNA in organisms

The amount of DNA recorded is the haploid content per cell. An amoeba has about a million times as much DNA as some bacteria, and a lungfish has about a hundred times as much as a human.

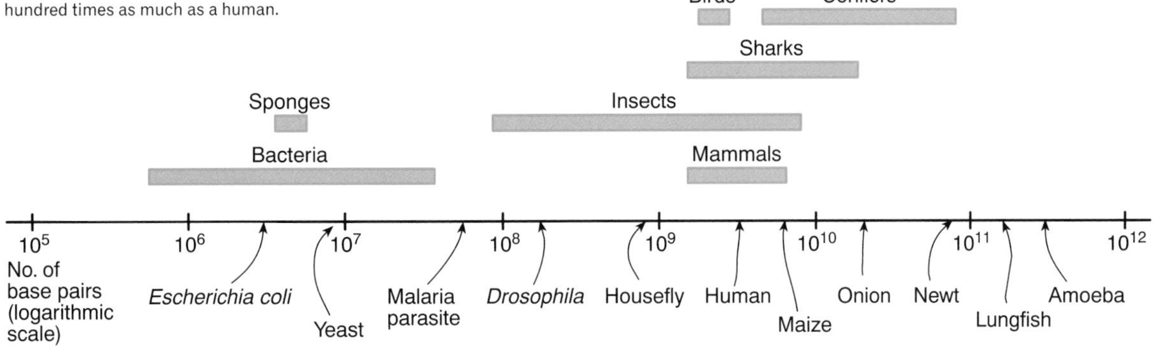

35

6.4 Summary

In this chapter three types of mutation have been described – polyploidy, chromosome mutations and point mutations. Mutations are due to imperfect copying of DNA or to imperfect chromosome separation. All mutations are therefore accidental and, like all accidents, their incidence is random and subject to chance. Polyploidy – doubling of chromosome number – is an important means of species formation in plants, but is rare in animals.

The various kinds of chromosome mutation lead to rearrangements of the genetic material, and have varying effects – from lethal to undetectable. They are an important cause of sterility in hybrids between species, tending to keep existing species distinct. One class of chromosome mutation – duplication – is a source of evolutionary novelty (Section 13.3).

Point mutations, changes in single 'letters' of the genetic code, are the most common type of mutation and produce many variant genes within a species. This situation recalls polymorphism, discussed in Section 3.3, and the multiple alleles discussed in Section 5.3, where members of a species may carry one of several variant genes at the same point on the chromosome. But to take the next step in the argument and assume that point mutations are the cause of multiple alleles is to make a guess about the past: that mutations arising in earlier members of the species have become widespread. This idea, of changes in gene frequency through time, introduces **natural selection**, which is discussed in the next two chapters.

Natural selection theory

Natural Selection is at the same time the explanation of everything and nothing in the Darwinian theory of evolution ... In all my life, I never have met a text which offers so many difficulties of all sorts as The Origin of Species.

Leon Croizat, *Space, Time, Form: the Biological Synthesis,* 1962

Natural selection – 'the struggle for existence', 'the preservation of favoured races', 'the survival of the fittest' – was the central idea of Darwin's theory of evolution. Darwin based his case on the discrepancy between the actual and potential offspring of an organism (or a pair of organisms in sexual species), on the vast destruction of gametes and young individuals entailed by this discrepancy, on the idea that some variations might be advantageous and others harmful, and on analogy with artificial selection, by human breeders, in domesticated animals and plants. The example Darwin used was not an obvious one like the female cod, capable of laying three or four million eggs in a year, but the slowest breeder of all – the elephant. Darwin calculated that the offspring of a single pair of elephants could number at least 15 million after 500 years. Since Darwin's work the lapse of time has been sufficient for many examples of natural selection to have been observed, both in nature and in laboratory experiments, and knowledge of genetics has given selection theory a sound basis. Unfortunately, selection theory (or theoretical population genetics) is now one of the most sophisticated and mathematical branches of biology, difficult to summarize or discuss in simple terms. It is fair to say that when he introduced natural selection, Darwin did not fully appreciate the subtlety of his idea. Natural selection may have appeared to the Victorians as a blunt instrument ('nature red in tooth and claw', in Tennyson's pre-Darwinian phrase) but in the modern theory it is an agent of remarkable refinement and variety, and some of those ideas must be introduced here, if only to give an idea of the complications that arise in understanding or explaining real situations.

In this chapter a theoretical example of selection is discussed, to bring in some aspects of population genetics. Chapter 8 discusses instances of selection in action.

Suppose that a new point mutation, involving a change in a single base pair, has appeared in one individual of a sexually reproducing species. What will be the fate of this mutant molecule? This depends first on where (i.e. in what cell) the mutation arises. If it is in a cell of the body the mutation will die with that individual and be lost, regardless of its effects. But if it arises in a reproductive cell (sperm, egg), or in a cell that will give rise to some or all* of the reproductive cells, there is a possibility that the mutation will be passed on to the next generation. If this is so, the mutant gene becomes interesting, and we can ask what its effects are.

There are three possibilities: that the effects are harmful, that they are neutral, or that they are beneficial. Most point mutations with detectable effects change the coding for one subunit of a protein molecule (page 30), and may affect the functioning of that molecule. Many such changes will be for the worse, some (for example, a large proportion of the known point mutations in human haemoglobin) will have no obvious effect, and a few may improve the functioning. It is easy to think of beneficial mutations in science fiction terms – wings sprouting from a mammal, or eyes appearing in the back of a man's head – but most beneficial mutations will be on a very mundane level, perhaps allowing a chemical reaction to proceed slightly faster, or at a slightly lower temperature.

Later we shall consider the fate of each of these three types of mutation but first it is necessary to ask how we

---

Even if the mutation arises in an embryonic cell that will, by division, give rise to all the gametes of the organism, the mutant DNA will be present only in half the gametes and will, on average, be transmitted to half the offspring. This is because the mutation will be in one chromosome of a homologous pair and half the gametes will receive the mutation-bearing chromosome, and half the normal one.

might recognize them. It should be evident that this can be done only by comparing the performance or success of the organisms carrying the mutation with those carrying the normal, non-mutant ('wild-type') gene. Success in this context is reproductive success, the efficiency with which the gene is transmitted to subsequent generations. The status of a mutation, whether it is advantageous, neutral or harmful, can be assessed only by observing the success of individuals carrying it, preferably over several generations. Instead of 'success' it is customary to use the Darwinian term 'fitness' ('adaptive value'), and to attach a numerical value to it. Such a value – an estimate of the fitness of a genotype – will always be relative, because one can only compare the fitness of one genotype with that of another, and in a particular environment.

Fitness in this technical sense is defined as an estimate of the contribution that a particular genotype will make to the next generation, expressed as a proportion of the gametes contributing to that generation.

Imagine that in a breeding population of a species a certain gene is represented in two alternative forms (alleles), symbolized by 'A' and 'a', and the proportion of genotypes is 25% AA, 50% Aa and 25% aa (Fig. 7.1). The proportion of each of the two alleles here is 50% (25% + 0.5 x 50%). This is also the ratio of the gametes that combine to produce this population – and we would expect the same proportions in the next generation. But suppose that in the next generation the genotypes are 36% AA, 48% Aa and 16% aa. The proportions of the two alleles are now 60% 'A' (36% + 0.5 x 48%) and 40% 'a' (16% + 0.5 x 48): the 'A' gene has been transmitted more efficiently than 'a' (whose effects are obviously relatively defective at some stage in the life cycle). For every 100 'A' gametes contributing to the first generation 120 contribute to the second, while for every 100 'a' gametes in the first generation there are only 80 in the second. The relative fitness of the two genotypes can be expressed by the ratio between these two numbers: for 'A' it is 120/100 or 1.2, and for 'a' 80/100 or 0.8. The mean fitness of the population is therefore

$$\frac{1.2 + 0.8}{2} = 1.0$$

The **selection coefficient** is defined as 1 – fitness. In this example for 'a' it will be:

$$1 - 0.8 = 0.2;$$

and for 'A':

$$1 - 1.2 = -0.2.$$

A positive selection coefficient means that a gene is decreasing in frequency in successive generations, and is relatively harmful, while a negative selection coefficient means that a gene is beneficial, and will increase in successive generations.

7.1   Harmful mutations

We have assumed that a harmful mutation appears in the reproductive cells of one individual of a population, let us say in a male. Some of the sperm produced by this individual will carry the mutant DNA molecule, and some may fertilize eggs produced by one or more females that lack the mutation, having the normal or 'wild-type' gene. The resulting fertilized eggs, and the individuals that develop from them, will be **heterozygous** for the mutation, carrying the mutant in one chromosome of a pair (the one from the father). Whether the mutation produces any effect in the heterozygotes will depend on whether it is dominant or recessive. A dominant mutant will affect the organism (the phenotype) in heterozygotes. A recessive one will be manifested only in **homozygotes**, which possess a Mendelian double-dose of the gene (see Section 4.1) – when it is the same in both chromosomes of the pair. Homozygous individuals must inherit the same form of the gene from both parents, so there can be no individuals homozygous for the new mutation until it is sufficiently widespread for two carriers to mate.

We do not always know why some genes are expressed in a single dose (dominant), why some are expressed only in homozygotes (recessive) and why some pairs of genes (alleles) are both active (incomplete dominance), but we do know that these terms are relative (page 23), depending on whether the visible effects or the protein product of the gene are considered. Nor can we always predict whether a new mutation will be dominant, recessive or incompletely dominant (in which case the organism will show the effects of both the normal and mutant genes). Most observed mutations are harmful and recessive but it does not follow that new mutations will also be recessive, for it is very likely that the mutations observed (in *Drosophila* or humans, for example) have appeared millions of times before in previous generations and it may be that selection has adjusted the dominant/recessive mechanism so that the mutation does least damage, and is recessive. We know that the dominant or recessive status of a particular gene is not fixed for all time, but can be changed by modifying genes. We also know that in *Drosophila*, for example, whenever a variety of mutant forms of a particular gene exist the mutants show incomplete dominance to one another but are recessive to the wild-type or normal gene. The most reasonable assumption is that new mutations will be incompletely dominant, manifesting some effect in heterozygotes.

The fate of a new harmful mutation will certainly be influenced by whether it is dominant, recessive or incompletely dominant. The effects of a dominant mutant will be fully expressed in the next generation; a partially dominant mutation will have some immediate effect in this generation; and a recessive mutation will not be expressed at all in

the heterozygotes. Its fate will also depend on how harmful it is – in population genetic terms, its fitness. A dominant lethal mutation that kills all carriers early in life, before they can reproduce, will disappear immediately, will have fitness zero and a selection coefficient (1 – fitness) of 1. A less harmful mutation might have greater fitness, perhaps 0.95, and a selection coefficient of 0.05: this would mean that if 100 individuals carried the mutation in one generation, only 95 in the next would do so. But it must be emphasized that all such numbers are statistical or probabilistic estimates and can be calculated only for large

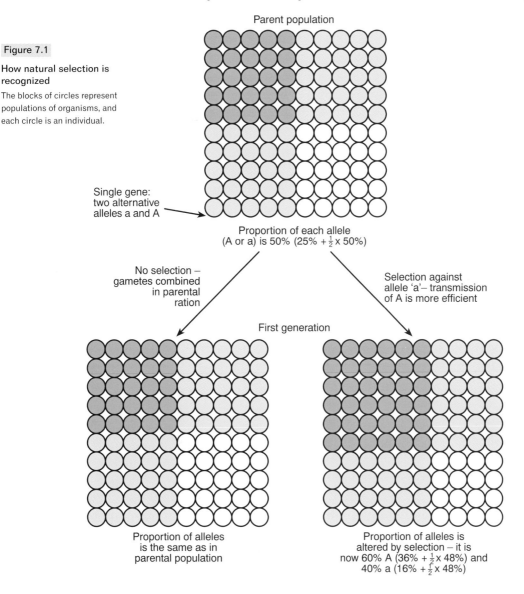

Figure 7.1

How natural selection is recognized

The blocks of circles represent populations of organisms, and each circle is an individual.

Parent population

Single gene: two alternative alleles a and A

Proportion of each allele (A or a) is 50% (25% + $\frac{1}{2}$ x 50%)

No selection – gametes combined in parental ration

Selection against allele 'a'– transmission of A is more efficient

First generation

Proportion of alleles is the same as in parental population

Proportion of alleles is altered by selection – it is now 60% A (36% + $\frac{1}{2}$ x 48%) and 40% a (16% + $\frac{1}{2}$ x 48%)

populations, in which the genes in question are represented in such large numbers that fluctuations due to chance can be neglected. In any particular situation these numbers no more predict the outcome than a bookmaker's odds predict the order in which racehorses pass the post. In the example above, where a new mutation in a single individual has been transmitted to a few progeny, it is quite possible that a predator will eat all the offspring at once and the mutation will be lost. But, neglecting such accidents, the probability of a mutation being transmitted to subsequent generations can be calculated, and whenever the gene confers some disadvantage (positive selection coefficient) it will eventually be eliminated. A dominant mutant, expressed in all its carriers, will be eliminated rapidly. An incompletely dominant gene, partially expressed in heterozygotes and fully expressed in homozygotes, will be eliminated more slowly. And a recessive gene, expressed only in homozygotes, will be eliminated very slowly indeed, for selection cannot act on it until it has become sufficiently widespread for matings between heterozygotes to occur (homozygotes are produced only from parents that both carry the gene). Box 7.1 discusses a well known example of the elimination of a harmful mutation, one that is unusual because it is inherited on the X-chromasome.

Some idea of the different rates of elimination comes from calculations of the 'force' tending to eliminate a disadvantageous gene from a population: this can be thought of as the number of generations necessary to reduce the frequency of the gene by a certain amount. For example, we can compare a recessive gene which confers a 1% disadvantage (in homozygotes) with an incompletely dominant gene with a 1% disadvantage in heterozygotes and a 2% disadvantage in homozygotes. If we imagine that these genes have become widespread in a population, to reduce the number of carriers of the incompletely dominant gene from one individual in 100 to one in 1000 will require about 230 generations, and the same number of generations will reduce the frequency from one in 1000 to one in 10 000. But with the recessive gene it will take about 90 000 generations to reduce carriers from one in 100 to one in 1000, and 900 000 generations to reduce it further to one in 10 000.

The reason for this enormous difference is that natural selection acts on individual organisms, eliminating inefficient phenotypes. But what is selected, and allowed to pass from generation to generation, is not the phenotype but the genotype. Recessive genes, not expressed in the phenotype

of heterozygotes, are to a large extent shielded from natural selection, especially if they are rare. This is because homozygous individuals, in whom the characteristic is expressed, have to receive the gene from both parents and will be very uncommon. Even if a recessive mutation is lethal, and kills all homozygotes early in life, it will take about 900 generations to reduce its frequency from one in 100 to one in 1000 of the population. This may seem to be useless knowledge, but it can have practical consequences. There are many different lethal and sublethal recessive mutations in the human population. It might seem a good idea to try to identify carriers (heterozygotes) of these genes and to discourage marriages between them, so avoiding the birth of afflicted homozygotes. But such a policy would actually increase the frequency of these harmful genes because they can be reduced in number only by the birth, and selective elimination, of homozygotes, or by preventing carriers from breeding at all.

Natural selection will eliminate harmful mutations from a population at a rate which varies with the harmfulness of the mutation (the lack of fitness of its carriers) and with its degree of dominance. This type of selection tends to maintain a species unchanged, and is called **stabilizing** or **purifying** selection.

7.2   Neutral mutations

If a neutral mutation, having no selective advantage or disadvantage, appears in one member of a population the number of individuals carrying it in subsequent generations will increase or decrease entirely by chance. The probability of various outcomes can be calculated, but here the analogy with a bookmaker's odds or a roulette wheel is more obvious. To understand how changes in gene frequency caused by chance alone, by **genetic drift**, may have a certain inevitability it may help to consider Figs 7.3 and 7.4.

The simplest conceivable situation is reproduction by fission with constant population size, shown in Fig. 7.3. If the population is to remain constant, only half the individuals in each generation may reproduce. In this simulation the five individuals that reproduce in each generation are chosen at random; after seven generations, the entire population is descended from one (coloured) individual in the first generation. If that individual carried a new mutation, the mutation would be fixed (universal) by the seventh generation, purely by chance. The point that may not be immediately obvious is that eventually, in *some* future generation, all the individuals in the population *must* be descended from

Box 7.1

Haemophilia in the descendants of Queen Victoria

Haemophilia is due to a mutation in one of the genes producing the proteins that cause clotting of the blood. Haemophiliacs (or 'bleeders') are at risk because they may bleed to death from trivial accidents. Haemophilia shows a peculiar pattern of inheritance, called sex-linked because the gene is on the X chromosome (see Fig. 4.3). The mutation is recessive and has no effect in females (XX), except in very rare cases where both parents carry the mutation. In males (XY) the mutation behaves as if it were dominant because there is only one X chromosome, one copy of the gene, and no 'wild-type' gene to mask its effects.

Queen Victoria (I 1) carried a new haemophilia mutant on one of her X chromosomes. Half of her children would be expected to carry the mutation, but only the male carriers would be haemophiliacs. The pattern of inheritance in her descendants is shown in Fig. 7.2. In fact, two of her five daughters were definite carriers. Her eldest daughter, Victoria (II 1), may have been a carrier, and there is no way of telling whether Louise (II 6), who died without offspring, was affected. Only one of her four sons, Leopold (II 8), was affected. Although frequently ill with internal bleeding, he lived 31 years before dying from a brain haemorrhage, and had two children, a son who was unaffected and a daughter who was a carrier (daughters of male haemophiliacs are always carriers, for they must have the father's X chromosome).

The daughter was Princess Alice (1883–1981), the longest surviving granddaughter of Queen Victoria, and probably the longest surviving carrier of the mutant gene. Princess Alice (III 22) had two sons – one who was haemophiliac and died at 21 (IV 28) and one who died in infancy (IV 29) and may have been – and a daughter (IV 27) whose descendants are free of the mutation.

Queen Victoria's granddaughter Alexandra (III 15) married Tsar Nicholas II and introduced haemophilia into the Russian royal family. Her son, the Tsarevitch (IV 26), was haemophiliac and her four daughters may have included carriers, but this branch became extinct when the family was assassinated in the Russian revolution.

Through Queen Victoria's youngest daughter the mutation was introduced into the Spanish royal family, but it is probably now extinct there. Through Queen Victoria's eldest daughter the mutation may have been introduced into the German royal family (IV 4, 6, 7), but it is now extinct there.

With Princess Alice's death in 1981 the harmful mutation that first occurred in Queen Victoria is assumed to have become extinct, and the pedigree terminates with the generation including her great-grandchildren and the present queen's children (VI 1-4); these children could not be carriers because both parents are descended from men who were not haemophiliacs.

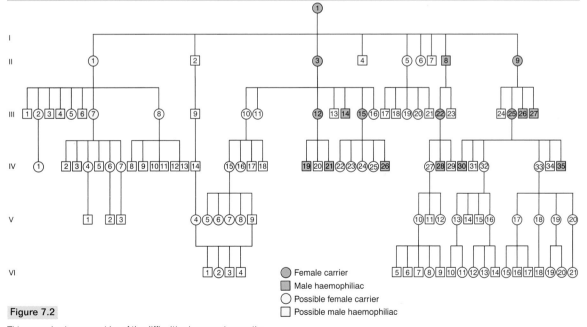

Figure 7.2

Female carrier
Male haemophiliac
Possible female carrier
Possible male haemophiliac

This example gives some idea of the difficulties in accurate genetic analysis extended over several generations. Even in genealogies as well known as those of European royalty there are many unanswerable questions, especially with a trait like haemophilia, where female carriers are identifiable only if they have affected male descendants. It is often impossible to establish the cause of deaths in infancy a century ago (e.g. III 4, III 6).

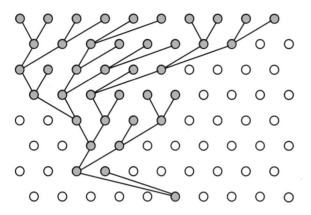

Figure 7.3

The fate of new mutations

This figure shows the simplest possible situation –
reproduction by asexual fission (as in bacteria and amoeba)
and with constant population size, here ten per generation.

just one member of the first generation, purely by chance.
If there is no selection, each of the individuals that reproduces in the first generation has an equal chance of beginning the successful lineage that will replace the others.

The situation in sexual reproduction, shown in Fig. 7.4,
is more intricate. In this example, four pairs, selected at
random, reproduce in each generation. Brother–sister matings are not permitted and population size is kept constant
at ten, with the number of offspring of each pair randomly
selected at 1–4. A new mutation is assumed to originate on
one chromosome of one individual (coloured) in the first
generation. On average, half the descendants of that individual will carry the mutation (as heterozygotes) in each
generation. In this simulation there is one carrier in the second generation, two in the third, three in the fourth, four
in the fifth, and the first homozygote (descended from two
carriers) occurs in the sixth generation. All the offspring of
a homozygote will carry the mutation, and if the homozygote mates with a carrier (heterozygote), half the offspring
will be homozygotes (two out of three in this example, in
the seventh generation).

The final generation in Fig. 7.4 includes seven heterozygotes and one homozygote, so that the mutation is present
in nine out of 20 chromosomes in that generation.
However, whereas in the asexual example (Fig. 7.3) all
the members of the final generation are descended from
just one member of the first generation, in Fig. 7.4 the

members of the final generation are each descended from
eight different members of the first generation.

The main difference between sexuality and lack of it
from the gene's view is that in simple asexually reproducing
organisms like bacteria there is only one copy of the gene
per individual so that a mutant gene is passed to all the
bearer's progeny, whereas in sexual organisms there are
two copies of the gene per cell (one on the maternal chromosome, one on the paternal) and a newly arisen mutant
will, on average, be passed to only half the bearer's progeny.
Nevertheless, the overall pattern of descent of a gene is
basically the same in sexual and asexual species. If all the
individuals are equally fit, chance alone will ensure that just
one of the copies of the gene in the first generation will
eventually be ancestral to every copy of the gene in a
descendent population.

A newly arisen neutral mutation is more likely to
become extinct than to father a lineage. In large sexually
reproducing populations, the odds are about 2:1 that the
new mutant gene will survive into the next generation,
about 6:1 that it will have disappeared by the tenth generation, about 50:1 that it will be gone by the hundredth generation and there is only one chance in 1000 that it will
persist for 1000 generations. However, these are probabilities, not predictions. Over long periods of time thousands
of neutral mutations might appear in a population, and
purely by chance some will increase in number sufficiently
to become widespread or even universal. Because we can

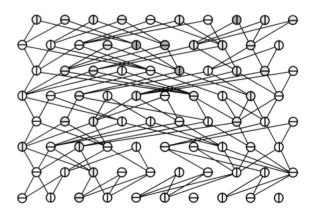

◐ Male
⊖ Female
◯ Heterozyote
● Homozygote

Figure 7.4

The fate of new mutations

This figure shows the more complicated
pattern of sexual reproduction.

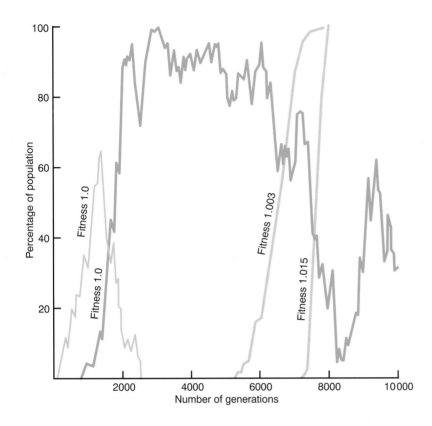

Figure 7.5

Computer simulations of the fate of new mutations

Some results when neutral (fitness = 1.0) and advantageous (fitness >1) mutations are simulated. The first mutation in the trial is neutral. It becomes widespread but then fades away and becomes extinct. The second is also neutral. It almost succeeds in becoming fixed (present in every individual), but continues to fluctuate and is present in only about one-third of the population at the end of the trial. The third and fourth mutations are advantageous. The third gives a very slight advantage (three in 1000), and becomes fixed after about 2000 generations. The fourth has greater fitness, and becomes fixed after 700 generations.

hardly ever observe new mutations, or their fate in the wild, computer simulations are the only way of visualizing and comparing the effects of selection and of chance (Fig. 7.5).

These probabilities apply in large populations, but increase in gene frequency by chance fluctuations is much more likely in small populations. To take the most extreme example, if the mutation arises in one member of a population of only two, such as Adam and Eve, the odds that it will be transmitted to half the next generation are even, and once a gene is present in half the population it is well on the way to being **fixed**, or present in all members. In nature, population bottlenecks like this, when a species or breeding population is reduced to only two members and then recovers, are extremely unlikely. But in a breeding population of 100 the chance of neutral mutations becoming fixed is far higher than in large populations: the average number of generations required to fix a mutation by chance is four times the effective breeding population – 400 generations for a population of 100, 4000 for one of 1000. This process, the chance fixation of a neutral (or even disadvantageous)

mutation in small populations, is an example of evolutionary change by genetic drift.

Experiments with the fly *Drosophila* have established that genetic drift can cause even slightly disadvantageous mutations to be fixed in small populations, against the pressure of selection. But the role of genetic drift in evolution is controversial, partly because of difficulties in recognizing neutral mutations or their effects. The fact that some attribute of an organism *appears* useless is no guarantee that it *is* useless. Investigation of apparently useless features has shown that some do, after all, have survival value. And while it may be possible to show that a certain feature has survival value and is subject to selection, it is never possible to prove that a feature has no survival value and is selectively neutral: we might not have looked carefully or closely enough.

Given our current knowledge of the organization of DNA and of the genetic code, neutral mutations surely occur. Any change in a base pair in DNA is a mutation. DNA consists of two main kinds of sequences of base pairs, **coding** sequences, triplets that are transcribed into RNA and then translated into protein, and **non-coding** sequences. In coding sequences **silent** or **synonymous** mutations can occur. Because the code is redundant, several different triplets coding for the same amino acid, many base pair changes (particularly in the third position of a triplet) will not alter the amino acid specified (for example, changing 'TAG' to 'TAT' – see page 28). These will not be detectable in the phenotype, and may be neutral (I write 'may' rather than 'must' because a different triplet, though coding the same amino acid, might require a different tRNA, perhaps a less abundant one, and so be disadvantageous). In non-coding DNA, such as the introns (page 21) or the short repeated sequences in minisatellites (page 27), base pair mutations will be neutral unless that position has some function in gene expression or control. However, no mutation that is expressed in the phenotype, causing some small difference in the constitution of the organism carrying it, can safely be regarded as neutral, with no selective advantage or disadvantage whatever.

7.3 Selection versus genetic drift

In the early days of genetics, before the nature of the hereditary material was known, it was easy to think of superficial features of organisms as direct expressions of single genes. Thus, one might imagine a 'gene for wrinkled seed-coat' in peas, or in *Drosophila* a gene for each pair of bristles that is subject to genetically controlled variation. Many species have features (such as individual bristles in insects, details of wing coloration in butterflies, or human fingerprint patterns) that are genetically controlled but which appear to be so trivial that they can make no difference to the survival of the individual and can have no selective value. Such features were a problem to Darwin and to later evolutionists, because these apparently useless details are often so characteristic for each species that we use them to name or identify organisms. If these features are useless, how did they become fixed by natural selection? Genetic drift – a mechanism which could fix neutral or useless features by chance – provided a welcome explanation of such difficulties.

However, as knowledge of genetics increased it became evident that there is no 'gene for third pair of bristles', or 'gene for fourteenth spot on wing': these features are the end products of an extremely complicated process (which we are still far from understanding fully) in which the coded information in the single nucleus of the fertilized egg is duplicated and progressively unfolded in the development of each individual. Although a pair of bristles in an insect may

be inherited as if it is controlled by a single gene, it is not the direct expression of that gene: the direct product of the gene is a protein or part of a protein, and the bristles are one result of a series of interactions between this protein and hundreds or thousands of others. With this view of the genetic control of development, it became reasonable to regard bristle or fingerprint patterns, apparently non-adaptive, as outward manifestations of genes which also made other, less obvious contributions, harmful or beneficial, and so are subject to selection. For a while most biologists believed that all aspects of an organism, no matter how useless they appeared, were shaped by natural selection.

In the last 30 years new stores of variation have been revealed. The first came from the amino acids in the proteins that are the direct expression of the genes. Variation here is studied most simply by electrophoresis, a technique which separates variant protein molecules by their response to an electric current. Almost every protein that is investigated is found to occur in several forms within a species (e.g. there are over 550 recorded variants of human haemoglobin; Section 6.1). Like the deviant haemoglobins, these variants, where they have been analysed, usually involve substitutions of a single amino acid. The more recent invention of methods of direct sequencing of DNA have revealed much more variation, the bulk of it at silent sites in coding sequences and in non-coding sequences. It becomes impossible to believe that all of these variants affect survival and can therefore be controlled by selection: many must be neutral mutations drifting towards extinction or fixation (Fig. 7.5). So in the last 20 years there has been a new surge of support for random ('non-Darwinian') evolution, regulated by genetic drift – chance fluctuations in gene frequency – rather than natural selection.

The question of whether the store of genetic variation found in protein and DNA studies is controlled mainly by selection or by random drift is the centre of the main controversy in evolutionary theory today. No molecular variant has yet been studied for long enough to decide whether its frequency is changing in a random way or not, and it is very doubtful whether long-term studies could resolve the question. The problem is that natural selection theory says that very small selection coefficients, of the order of 1% or less, are effective in causing evolutionary change, yet the demonstration of such small differences in fitness is simply not possible in experiments. It has been calculated that a 1% difference in fertility between two genotypes could be shown with 95% confidence only if the fertility of 130 000 females of each type were measured. If the fertility of 380 females of each type were measured, the investigator has only an even chance of detecting a much larger difference in fertility (10%). Selection theory is thus trapped in its own sophistication: it asserts that small differences in fitness are effective agents of evolutionary change, yet differences of that order are not detectable in practice.

The argument between the selectionists and those who assign an important role to random effects ('neutralists') is therefore largely confined to paper. The consequences of the two opposing theories are not sufficiently different for them to be discriminated by experiment. A further difficulty is that the two theories are not mutually exclusive. Advocates of natural selection do not assert that *all* change is guided by selection: they allow the possibility of neutral mutations and random events in small populations. And advocates of neutral mutation and genetic drift have never supposed that these are the cause of *all* evolution: they agree that selection plays an important part as the only mechanism we know that can cause adaptive change.

This question of random or neutral evolution is explored further in Chapters 9 and 10, but this brief account of the controversy is enough to show that it bears on an important philosophical point, the status of evolutionary theory as science. This point is explained more fully in Chapter 14, but it can be set out briefly here. Darwinian evolution, by natural selection, predicts that organisms are as they are because all their genes have been (and are being) subjected to selection – mutations that reduce the organism's success are eliminated, and those that enhance it favoured. This is a scientific theory, for these predictions can be tested. 'Non-Darwinian' (or neutral) evolution predicts that some features of organisms are non-adaptive, having neutral or slightly negative survival value, and that the genes controlling such features are fluctuating randomly in the population or have been fixed because at some time in the past the population went through a bottleneck, when it was greatly reduced. When these two theories are combined as a general explanation of evolutionary change, that general theory is no longer easily testable. Take natural selection: no matter how many cases fail to yield to a natural selection analysis, the theory is not threatened, for it can always be said that these failures of selection theory are explained by genetic drift. And no matter how many supposed examples of genetic drift are shown to be due, after all, to natural selection, the neutral theory is not threatened, for it never pretended to explain all evolution.

Perhaps the best we can do is to use genetic drift, as an 'explanation' of otherwise inexplicable facts, as sparingly as possible for, unless the effects of natural selection are looked for with the greatest care, they will not be found.

7.4 Favourable mutations

In nature, new favourable mutations must be rare. This is because existing species are the result of past selection, which will have brought them close to the best possible adaptation to their surroundings, so that mutations (changes) will decrease that adaptation. Favourable mutations are more likely in a changing or deteriorating environment, when mutations that were previously harmful may become valuable.

If a favourable mutation appears in one member of a population its fate, as in the case of harmful mutations, will be influenced by whether it is dominant, recessive, or incompletely dominant. But initially its fate is likely to be more subject to chance than is a harmful mutation. This is because beneficial mutations have a much narrower range than harmful mutations. Harmful mutations range from those with very slight effects to those which are lethal, reducing the fitness of organisms carrying them to zero. Favourable mutations with effects as drastic as lethal mutations would reduce the fitness of all other members of the species to zero – which is possible, but only in severely deteriorating environments. In stable environments favourable mutations are likely to have only small effects, and whether they increase in frequency or disappear will be much influenced by chance. The odds that a beneficial mutation will survive are very close to those for a neutral mutation. Table 7.1 compares the probabilities of survival of a neutral mutation (no advantage) and a mutation conferring a 1% advantage, each of which has appeared in one individual.

Table 7.1

Generation	Probability of survival	
	No advantage	1% advantage
1	0.632	0.636
3	0.374	0.380
7	0.209	0.217
15	0.113	0.122
31	0.059	0.069
63	0.030	0.041
127	0.015	0.027

The figures in Table 7.1 may be read as the number of cases per 1000 in which survival is probable. Thus after 15 generations the favourable mutation is likely to survive only in nine more trials in 1000 than the neutral mutation. Looked at in another way, a favourable mutation conferring a 1% advantage may have to occur about 50 times before, by chance, it becomes sufficiently widespread in the population to have a secure future.

Once a favourable mutation gains a secure foothold in the population, the rate at which it will spread and replace the non-mutant alternative depends largely on whether it is dominant or recessive, just as with a harmful mutation. A mutation giving a small advantage and one giving a small disadvantage are two sides of the same coin, for if an advantageous mutation appears, its alternative – the existing non-mutant gene – will thereafter have a corresponding disadvantage, and the spread of the advantageous mutation will eliminate the non-mutant gene. The figures illustrating the slow elimination of a harmful recessive (Section 7.1) can equally well be read as the final stages in the spread of an advantageous dominant. The different patterns of spread of a dominant or recessive favourable mutation, and elimination of a dominant or recessive harmful one are contrasted in Fig. 7.6. Once the advantageous mutation gains a foothold it will eventually replace its alternative, so that there will be a change in the genetic constitution of the population. This is called **directional** or **positive** selection.

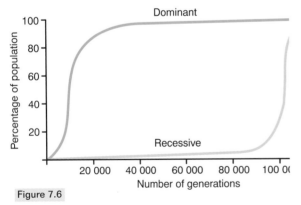

Figure 7.6

The spread of dominant and recessive mutations

A dominant mutation spreads through 90% of the population quite rapidly, but is still not fixed even after 100 000 generations. The recessive mutation spreads very slowly until it is present in about 10% of the population, and then becomes fixed quite quickly. These are ideal curves, showing the calculated rate of increase and neglecting chance fluctuations.

When considering advantageous mutations one type of dominance is particularly interesting. This is 'hybrid vigour' or 'overdominance', in which both the mutant and non-mutant homozygotes are less fit than the heterozygotes. Imagine a gene (say A_1) that codes a certain enzyme, and a partially dominant mutant form (A_2) that codes a variant protein with slightly different properties. The homozygotes (A_1A_1 or A_2A_2) will each have only one form of the enzyme, but the heterozygotes (A_1A_2) will have both, and might function more efficiently because of this. In such cases, although natural selection will tend to eliminate both types of homozygote, selection will not result in the elimination of either gene, but in a polymorphic population, with the two genes present in proportions that are governed by the relative disadvantage of each homozygote to the heterozygote. Even if one of the homozygous conditions is lethal, selection will preserve that gene. Suppose that A_2 is lethal in homozygotes (A_2A_2 has fitness zero) and A_1 homozygotes suffer a 5% disadvantage relative to the heterozygote (i.e. A_1A_1 has fitness 0.95; A_1A_2 has fitness one). After many generations, the result of this situation will be a stable population containing two or three lethal homozygotes per 1000 individuals, about 90 per 1000 heterozygotes, and the rest A_1A_1 homozygotes. If the two homozygotes are equally fit the result of selection will be a population containing 25% of each homozygote and 50% heterozygotes. Where there are three (or more) alternative forms of the gene the situation is more complicated, but selection will again produce a polymorphic population. This sort of selection, due to hybrid vigour and resulting in stable polymorphism, is called **balancing** selection. It is a type of selection that will preserve variation within a population.

7.5 Summary

This account of selection theory has been long, and probably heavy going. The theory is, in fact, even more complicated. In this chapter I have discussed selection of a point mutation as if each gene could be considered in isolation, but it cannot. Genes do not exist in isolation but in chromosomes and, although chromosomes are mixed up with their homologues by crossing over and can be broken up by chromosome mutations such as translocations, there is still strong linkage between adjacent genes on the chromosome. The prospects of a gene may thus be influenced by its neighbours – a neutral mutation might spread through the population by selection because it lies next to, and so is linked with, another mutant gene that is subject to strong selection. Natural selection theory also has to account for the spread of chromosome mutations – inversions, translocations, etc. (Section 6.2). For example, the eight inversions that distinguish human and chimpanzee chromosomes must each have originated in one individual in the ancestry of one species or the other, and spread through the population in the same way as a point mutation. Natural selection theory must consider chromosomes and pieces of chromosome, as well as individual genes.

It will be as well to summarize selection theory as briefly as possible. Mutations are random events, the result of accidental errors in replication of DNA or in nuclear division. A mutant DNA molecule or chromosome has no chance of propagating itself unless it arises in a sex cell or in a cell that will produce sex cells. Its fate will then depend on three factors:

- chance;
- whether a mutant gene is dominant, recessive or intermediate; and
- whether its effects are harmful, beneficial or neutral.

A mutation can be categorized as harmful or beneficial only after observing the performance of its carriers. If a mutation alters the phenotype its effects can never be proved to be completely neutral.

The terms 'harmful' and 'favourable' or 'beneficial' refer to reproductive success, and can be graded on a scale of fitness. This term is relative, because we can compare the fitness of one genotype only with another, and in a particular environment. In nature the great majority of mutations will be harmful – existing species, because of past selection, should already be closely adapted to their environments.

Chance is most important in the early stages of the career of a mutation, when it is present in only one or a few individuals, but it may also play a large part in small populations. Even in large populations all the existing copies of a gene must originate, even in the absence of selection, from some one ancestral copy. On average a harmful mutation will eventually be eliminated by poor reproductive performance of individuals carrying it, and the species will be preserved unchanged by stabilizing selection. An advantageous mutation will (on average) eventually become widespread by poor reproductive performance of individuals carrying the original, non-mutant, gene and the constitution of the species will change by directional selection. Balancing selection – when heterozygotes are fitter than either mutant or non-mutant homozygotes – results in polymorphism or the preservation of variation. Genetic drift, the chance fixation of neutral or harmful mutations in small populations, should be invoked as sparingly as possible.

8 Selection in action

This chapter will examine five examples of natural selection that have been observed and analysed during the last 50 years. They illustrate various aspects of selection theory described in the last chapter.

8.1 Sickle-cell anaemia

In Chapter 6 the numerous (over 550) known variants of human haemoglobin were mentioned (page 30). Most of these cause no obvious ill-effects in their carriers but one, known as haemoglobin-S, produces a severe disability – sickle-cell anaemia. This disease is so called because some of the red blood cells (erythrocytes) of afflicted people are sickle- or spindle-shaped, not disc-shaped as in normal blood (Fig. 8.1). Analysis shows that the only difference from normal erythrocytes is a single amino acid in the beta chain of haemoglobin – valine appears instead of glutamic acid at position six. It is therefore caused by a point mutation in DNA, substitution of 'A' for 'T' in the middle of the triplet, and is due to a mutant gene which we can symbolize as Hs (with Hn for the normal gene). Substitution of valine for glutamic acid alters the electric charge of the haemoglobin molecule – and because of this when the blood is deoxygenated the molecules tend to bond together in long fibres which cause sickling of the cells, blockage of capillaries by clumps of them, and a cascade of ensuing complications.

Sickle-cell anaemia is a severe disease and usually results in death before adolescence. Studies of its inheritance show that the severe, fatal anaemia is manifested only in homozygous individuals, who have received the mutant gene from both parents and have the genotype HsHs. In heterozygotes (HnHs), who receive the mutant gene from only one parent, the erythrocytes may show slight distortion when the oxygen concentration in the blood is very low, but these people are not anaemic and can lead normal lives.

Sickle-cell anaemia is found in three main areas:
- central Africa;
- in a zone extending through the eastern Mediterranean, Middle East and India to the Far East (Fig. 8.2);
- North and Central America.

Figure 8.1

Sickle-cell blood

A blood smear from a person heterozygous for the sickle-cell gene. Among the normal rounded cells are a few elongated cells, distorted by crystallization of the abnormal haemoglobin. Enlarged about 1200 times.

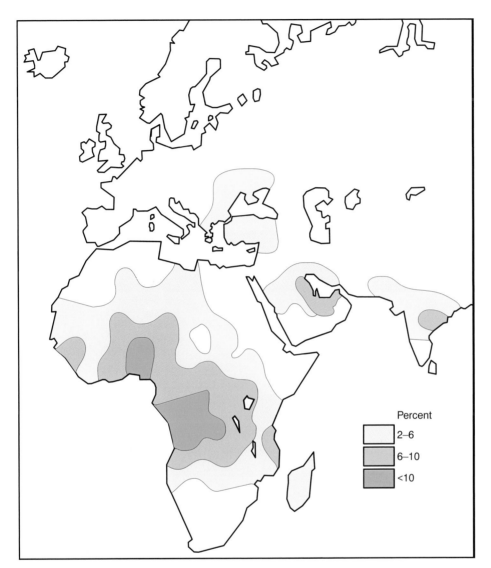

Figure 8.2

Sickle-cell gene

Distribution of the sickle-cell gene in the human population of the Old World. The incidence is highest in West Africa, which was the source of most of the slaves transported to the New World.

The key to this distribution was found in Africa, where the proportion of heterozygotes for the sickle gene is as high as 40% in some districts. These districts are those where the most severe form of malaria, malignant tertian or falciparum, is prevalent. Malaria is caused by minute, single-celled parasites that are introduced into the blood by the bite of a mosquito and then undergo growth and asexual reproduction cycles inside the erythrocytes, where they feed by breaking down haemoglobin. People with normal haemoglobin (genotype HnHn) are susceptible to this severe, and often fatal, disease. But those who are heterozygous for the sickle-cell trait (genotype HnHs) are much more resistant to malaria because infected cells tend to 'sickle' or collapse, interfering with the development of the

Figure 8.3

Peppered moth,
Biston betularia
On the left is a tree
encrusted with
lichens, from an
unpolluted rural area,
and on the right a
soot-covered tree from
an industrial city.
On each there is one
melanic moth and
one pale, typical form.

parasite. So, in areas where falciparum malaria is prevalent, sickle-cell heterozygotes are fitter than either the sickle-cell homozygote (who dies in childhood of anaemia) or the normal homozygote (who may die of malaria, also usually in childhood). The high proportion of sickle-cell heterozygotes in parts of Africa is therefore caused by balancing selection, where a mutant gene that is lethal in homozygotes is preserved by selection because of the increased fitness of heterozygotes in a malarial environment.

The slave trade gave a new twist to the sickle-cell story. This forced emigration introduced large numbers of Africans carrying the sickle-cell gene into America.

Through subsequent migrations the gene is now widespread in North and South America, and also occurs in Britain. In Central America there are malarial districts where the gene may be present in up to 20% of the population, but in North America there is little or no falciparum malaria, so the selective agent that maintained the polymorphism is no longer active. The average incidence of the sickle-cell gene in North American Blacks is less than 5%. In the absence of malaria, the sickle-cell gene is effectively recessive, heterozygotes having no obvious advantage or disadvantage over normal homozygotes. So in North America at present there is a situation like that outlined on page 40, where a lethal recessive is widespread, and where heterozygotes can be identified (by blood tests). Heterozygotes are advised not to marry another carrier of the gene but (as was shown on page 40) that will not reduce the frequency of the gene. Should these people be advised not to reproduce at all, to avoid propagating the gene? Not in a free society.

Sickle-cell anaemia illustrates several points:

- the possible lethal effects of a single point mutation;
- balancing selection, leading to a stable polymorphism in malarial environments;
- the fact that fitness is determined by the environment, the superior fitness of heterozygotes depending on whether they are in malarial areas or not;
- the difficult ethical decisions involved in genetic counselling.

8.2   Industrial melanism

The phenomenon of industrial melanism (blackening) has appeared in many insects during the last 150 years, as one consequence of the industrial revolution. In the neighbourhood of industrial towns and cities the buildings, rocks and tree trunks are blackened by deposits of soot, and the lichens and other simple plants that usually colonize such surfaces are killed. These changes have evoked a rapid and obvious evolutionary response in the British peppered moth, *Biston betularia*. This, like most moths, is nocturnal, flying by night and resting by day, generally in trees. The typical form of the moth has pale wings with small black markings, and on lichen-covered branches it is almost invisible (Fig. 8.3).

In all eighteenth- and early nineteenth-century insect collections, the peppered moth has this pale coloration. About 1850, melanic or black examples began to be caught near Manchester, and by 1900 almost all the peppered moths collected there were black. In the 1870s black individuals were still fairly uncommon but in the 1880s they outnumbered the pale form. In the Manchester district this change from a pale-coloured population of the peppered moth to a 98% dark population took about 50 years, and that period corresponds with the most rapid increase in the human population of Manchester and in the quantity of coal burned there. The same change in the peppered moth occurred around other British cities – black individuals were not recorded in London until 1897 but by 1915 they accounted for three-quarters of a large sample in a northern suburb – and the melanic form also became common in some nearby rural, unpolluted, areas. In North America the same thing happened with the closely related moth *Biston cognataria*. The melanic form was first recorded in Philadelphia in 1906, was found in Chicago in 1935 and constituted over 90% of the population in parts of Michigan by 1960. In Britain, smoke-control legislation was introduced in 1956, and pollution controls have since become gradually more stringent. The peppered moth has responded – at one carefully monitored site 15 km west of Liverpool the proportion of the melanic form was 92% in 1959, 61% in 1984, 30% in 1989 and less than 20% in 1994 (see Fig. 8.6); at Cambridge it was also over 90% in the 1950s and is now less than 30%.

The differences between the pale peppered moth (form *typica*) and the dark variety (*carbonaria*) is due to a single dominant mutation so that homozygotes and heterozygotes for the mutation are both black. There is also another, less common dark form (*insularia*) which is produced by at least three other mutations of the same gene, all dominant to *typica* but recessive to *carbonaria*. The very rapid spread of the dark form in industrial areas about 100 years ago and the equally rapid decline in the last 30 years are facts that demand explanation. Why should black moths be abundant in polluted areas, and pale ones in unpolluted (Fig. 8.4)? Two possible explanations have been offered:

- that pollutants are the direct cause;
- that they are the indirect cause, with natural selection as the direct cause.

The first explanation rests on a series of experiments in the 1920s and 1930s in which moth caterpillars of various species were fed on plants charged with pollutants, particularly lead and manganese salts. In certain experiments black or partially black individuals appeared in subsequent generations; some of these bred on as dominants, some as recessives and some as partially dominant. The explanation offered was induced mutation – that the pollutants were acting like X-rays or other mutagens and producing high mutation rates towards melanic forms. This is a 'Lamarckian' explanation (see page 149 for Lamarck), implying that the inheritance of acquired characteristics (or, as Lamarck put it, 'the power of new conditions and habits to modify the organs of living bodies') is a cause of evolutionary change.

The explanation for the spread of black moths by natural selection – a Darwinian explanation – rests on a series of experiments carried out during the 1950s. These showed that predation by birds is the cause. Peppered moths emerge from the chrysalis in the evening and fly during their first night of adult life. Females fly for about 30 minutes and then settle on a tree. They fly little on subsequent nights, and are likely to mate and deposit their eggs on the tree where they first settle. Males fly more actively, up to 5 km per night, but when one finds a female the mating pair remain in copulation for about 24 hours, typically on the

underside of a branch. This mating period is probably the most hazardous part of the moth's adult life, and once the female has laid her eggs the pair have done their bit, leaving the next generation to take over as the target for natural selection. A mated pair of pale (*typica*) moths looks very much like a growth of lichen, and is well camouflaged on a lichen-covered branch. A pair of black (*carbonaria*) moths will be conspicuous on such a branch, but inconspicuous on polluted trees (see Fig. 8.3). A mixed pair will be conspicuous in either environment.

In experiments large numbers of male, laboratory-bred pale and black moths were marked with a spot of paint on the underside. One sample was released in an unpolluted rural wood where *carbonaria* did not occur, and other samples were released in a polluted urban wood where *carbonaria* was abundant. Each night, light traps or traps 'baited' with a female were set and the number of marked

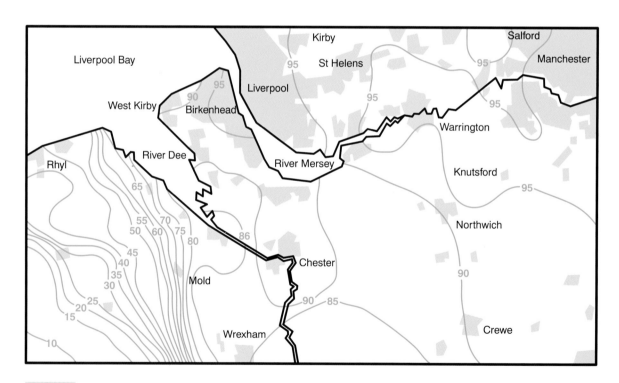

Figure 8.4

Distribution of light and dark peppered moths around Liverpool, England, in the early 1970s

The contours show the percentage of dark forms in the moth population: they were very common in the urban and industrial areas to the north and east but their numbers fell away rapidly towards the south-west, in rural Wales. The proportion of dark moths has since declined sharply (see Fig. 8.6).

moths trapped was recorded. In the rural wood 13% of the pale forms and 6% of the dark ones were recaptured, whereas in the urban wood those proportions were reversed, with 13% pale forms and 28% dark coming to the traps. That birds selectively eat those moths which contrast with their background was demonstrated in aviaries and in the woods during the experiments. Given these facts, it is reasonable to argue that the selective agent that caused the rapid spread of the dark form in industrial areas, and virtual elimination of the pale form, is predation by birds. Melanic

mutations have presumably been turning up spontaneously for thousands of years, but the mutant moths would be conspicuous in their usual resting places – lichen-covered trees – and would be seen and eaten by birds. With the advent of industrial pollution, the mutant form was suddenly at an advantage because it is camouflaged on blackened trees where the pale *typica* is highly conspicuous.

Which of the two explanations is correct, the Lamarckian one of direct effect of pollutants as mutagens or the Darwinian one of differential predation? The

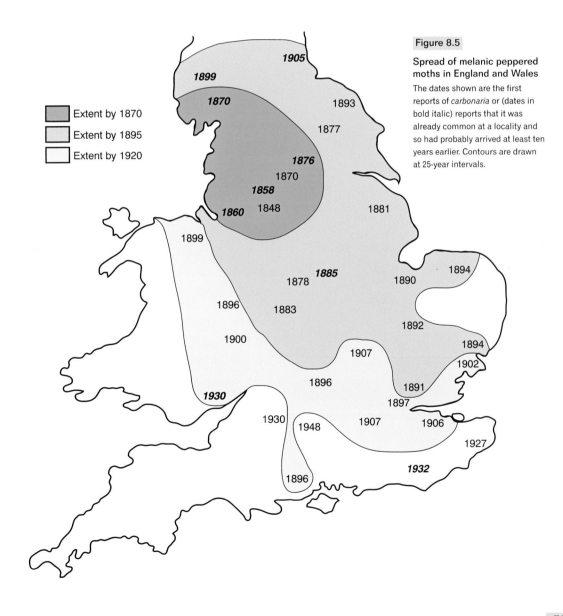

Figure 8.5

Spread of melanic peppered moths in England and Wales

The dates shown are the first reports of *carbonaria* or (dates in bold italic) reports that it was already common at a locality and so had probably arrived at least ten years earlier. Contours are drawn at 25-year intervals.

Extent by 1870
Extent by 1895
Extent by 1920

Lamarckian explanation has at least three disadvantages: lack of a general mechanism by which acquired characteristics can be transferred into DNA, the fact that the experiments showing induction of melanic forms by pollutants have so far proved unrepeatable, and the timing of the spread of melanic peppered moths in Britain (Fig. 8.5). If the melanic moths were directly produced by pollutants, we should expect them to appear wherever intense industrial pollution exists, but the map in Fig. 8.5 shows an obvious time lag between the appearance around Manchester and the appearance in other centres of industry in the north-east and south of England. The same kind of time lag has been seen with the spread of melanic *Biston cognataria* in North America.

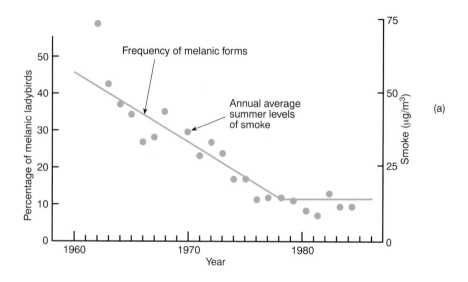

(a)

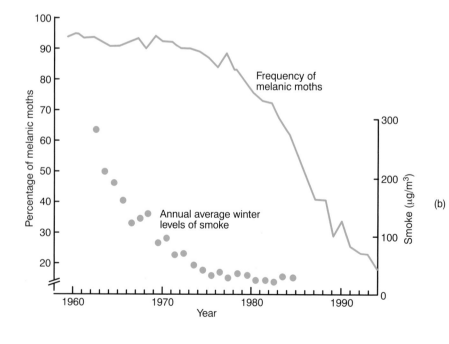

(b)

Figure 8.6

Changes in air pollution and frequency of melanic peppered moths and two-spot ladybirds

(a) The frequency of the two-spot ladybird in Birmingham correlates closely with levels of smoke.

(b) The peppered moth at West Kirby, about 10 km west of Liverpool. As in Birmingham, pollution had declined sharply by the middle 1970s, but the decline in dark moths lags well behind the decline in pollution.

The Darwinian explanation has been criticized for two reasons. First, the experiments supposedly demonstrating differential predation by birds were artificial, presenting birds with a glut of a normally rare food (a typical density of peppered moths is 15/km²) and with most of the moths in unnatural sites on tree trunks where they are easily visible, rather than in their preferred position beneath branches in the canopy. Second, subsequent experiments have failed to reveal such strong selection and a few have found none at all. Nevertheless, birds do eat the moths and there is surely selection against black moths in unpolluted woods and against pale moths in cities. Yet parts of Britain (like rural East Anglia) appear to be unpolluted but have up to 80% melanic peppered moths, and even in the most polluted environments there is never a complete take-over by melanic forms, so that differential predation by birds can hardly be the complete answer.

So far, the best explanation for the observed distribution of the different forms of moth is that predation of poorly camouflaged adults is supplemented by the black *carbonaria* having an advantage of about 20% in larval stages. In rural areas the conspicuous black adults are at a disadvantage but their numbers are supplemented because of the higher survival rate of *carbonaria* caterpillars. One could also explain the abundance of adult *carbonaria* in East Anglia by the fact that newly hatched caterpillars suspend themselves on silk threads, to be dispersed by the wind, and prevailing winds blow from polluted London into East Anglia.

There are other ways in which the problem can be approached. Moths are by no means the only animals to show industrial melanism. The two-spot ladybird, *Adalia bipunctata*, is a small beetle in which a black form is dominant to the usual red one. Ladybirds are conspicuous insects, protected from predators not by camouflage but by being unpalatable. Like black peppered moths, black two-spot ladybirds dramatically increased in many industrial areas in Britain and began to decline about 1960. Camouflage from predators cannot be the explanation here, and the story seems to involve something much more direct – the greater efficiency of black objects in absorbing heat. In an atmosphere darkened by smoke, a black ladybird will be able to warm up in spring earlier than a red one, and so may get a head start in emerging from hibernation and beginning feeding and reproduction. The decline in numbers of black peppered moths and two-spot ladybirds at two sites is shown in Fig. 8.6. The black ladybirds became more rare during the 1960s and 1970s, in harmony with

reduction in smoke, but the decline in black peppered moths lags behind the fall in smoke by about 10 years. This is what we would expect if pollution were controlling the ladybirds directly, by blocking out the sun, and controlling the moths indirectly, by killing off the lichens among which they are concealed, because it will take some years for lichen colonies to re-establish after pollution falls below levels lethal to them.

Industrial melanism brings out several aspects of natural selection. Perhaps the most important is the idea that directional selection, leading to a change in the gene pool of a species, is caused by changes in the environment. It is also obvious that the fitness of any mutation is determined by the environment – the black mutation is harmful in moths or ladybirds in rural areas but advantageous in industrial areas. However, the story of industrial melanism seems nowhere near as simple as it did 20 years ago, when the peppered moth was accepted as the classic case of natural selection in action, or more graphically as 'Darwin's missing evidence.' Reduction in urban and industrial air pollution, which regrettably still has far to go, provides a wonderful opportunity for detailed monitoring of the response amongst the numerous (over 200) species of insects and other animals that show industrial melanism. In the peppered moth, the most thoroughly studied example, increased knowledge has decreased the likelihood that predation by birds is the whole answer but has strengthened the case for natural selection as the cause.

8.3 Resistance to antibiotics

Antibiotics are substances, naturally produced by some micro-organisms, which inhibit the growth of other micro-organisms when present in extremely small quantities. The best known antibiotic, penicillin, is produced by the mould *Penicillium chrysogenum* and is active against a wide range of disease-producing bacteria. Penicillin was discovered in 1928, and was isolated and produced commercially during the Second World War. Since then, more than 1000 similar substances have been found, and many have come into commercial production – actinomycin, ampicillin, aureomycin, neomycin, streptomycin, tetracyclines and other familiar names. This multiplication is not due just to competition between drug companies – it is a response to evolution of resistance, by natural selection, in disease-producing bacteria.

Under favourable conditions bacteria may divide (reproduce) every 20 or 30 minutes, and a person suffering from a bacterial infection may harbour a population of bacteria many times greater than the human population of the world. If treated with an antibiotic an infected person might make a rapid recovery, but to the bacteria the antibiotic appears as a catastrophic change in a previously benign environment. For example, penicillin acts by preventing the formation of the bacterial cell wall, stopping cell division (other antibiotics inhibit protein synthesis or damage the surface membrane of the bacteria). But in such enormous populations the possibility of a spontaneous mutation giving resistance to the antibiotic may be quite high. The human colon bacillus *Escherichia coli* has a resistance mutation rate to streptomycin of between one in a hundred million and one in a thousand million cell divisions. Since a single bacterium may produce ten thousand million descendants within 24 hours, these odds become quite low, and resistance to streptomycin and other antibiotics is common. Some of these resistant strains turn out to be streptomycin dependent, and cannot live without the antibiotic, but back mutations to the original constitution occur at about the same rate, so that a two-mutation process – extraordinarily unlikely in one individual, but quite probable in a population numbered in billions – will enable a bacterial strain to survive attack by streptomycin unchanged.

Antibiotic-resistant bacteria show the maximum possible rate of evolution by natural selection – all other members of the population are unable to grow and have fitness zero, while the resistant mutant has fitness one, and replaces the non-resistant form almost instantaneously. The genetic systems of bacteria are much simpler than those of moths or humans: bacteria reproduce asexually by fission and are haploid, inheriting only one set of genes, so that any new mutation is expressed and there are no complications such as dominant and recessive forms of a gene. Although bacteria have no true sexual reproduction, there are ways in which DNA can be transmitted within and between bacterial species, and since these have been important both in the evolution of antibiotic resistance and in the development of genetic engineering (which is a kind of man-made evolution) they deserve some mention.

The two most important ways of transmitting DNA between bacteria are **transduction**, where the transmitting agents are viruses, and **conjugation**, where the agents are **plasmids**. Viruses are hardly more than bits of DNA (or RNA), a few genes enclosed in a protein coat. They can reproduce only by using the genetic system of an infected organism, after the viral DNA has been injected into the host cell. Bacteria are infected by viruses, just as we are, but

bacteria also contain more benign bits of DNA called plasmids. The difference between bacterial viruses and plasmids is easily understood in terms of human association with bacteria. We may suffer infection by various pathogenic (disease-producing) bacteria, such as *Salmonella* or those causing gonorrhoea or dysentery. We are all also infected with benign or beneficial bacteria such as the intestinal bacillus *E. coli*, with which we share a symbiotic relationship: *E. coli* synthesizes vitamin K, which benefits humans; we supply them with a constant food supply in a protected environment. Plasmids act as symbionts within bacteria. Each plasmid consists of a closed ring of DNA which uses the host cell's genetic system to reproduce and to translate its genes. Plasmids probably evolved from viruses that have achieved a balance with their hosts, just as our *E. coli* are probably descended from originally pathogenic bacteria.

One of the best known ways in which plasmids may benefit their bacterial hosts is by conferring resistance to antibiotics. By accident, both plasmids and bacterial viruses may pick up bits of their host's chromosomal DNA and incorporate it in their own. These accidents, which are a form of mutation, will be propagated by natural selection if they give some advantage, and they may spread between strains or species of bacteria when viruses infect new hosts and plasmids are transmitted in bacterial conjugation. None of the plasmids in bacterial strains isolated before 1940 contain genes resistant to antibiotics, but such genes appear in plasmids isolated during the 1950s – and some found today have genes resistant to four or five antibiotics. These bacterial symbionts obviously confer a great advantage to their hosts and therefore also benefit the plasmids, which would otherwise be exterminated when their hosts are killed off by an antibiotic. Multiple antibiotic resistance in pathogenic bacteria is generally due to plasmids rather than mutations in chromosomal DNA, and is obviously a great problem in disease control.

Genetic engineering, first pioneered during the 1970s, aims to introduce new DNA into lineages of organisms. Thinking only of bacteria, some examples of the purposes of such work include the 'manufacture' of bacteria that will synthesize human insulin or other hormones, will synthesize antibiotics or will feed on oil (to clear up oil spills). The techniques of genetic engineering depend on methods of isolating DNA, on enzymes that will snip and join DNA and on vectors that will introduce DNA into cells. Bacterial viruses and plasmids are used as vectors, and a most efficient method is to use a virus or plasmid containing an antibiotic

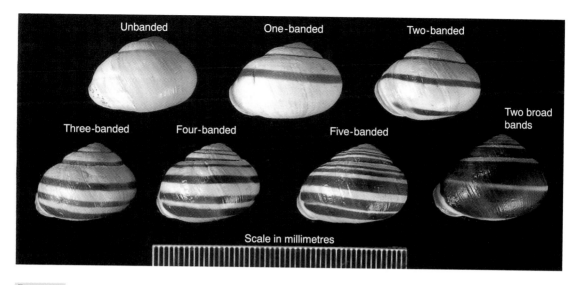

Figure 8.7

The snail, *Cepaea nemoralis*

Shells from a single population of *Cepaea nemoralis* in Cornwall,
England. The bands are counted on the largest whorl of the shell.

resistance gene. If the gene one wants to propagate can be
spliced into the DNA of such plasmids or viruses, a culture of
non-resistant bacteria can be exposed to the vectors and the
infected bacteria selected by growing them on a medium
containing the antibiotic, which will kill off all others.
Natural selection is being put to work in the laboratory.

Commercial production of penicillin is an example of
extremely successful artificial selection. The original strain of
Penicillium chrysogenum had a yield of penicillin that we can
call 1.0. A spontaneous mutation selected from the cultures
increased this yield to 2.5. Cultures subjected to X-rays produced a mutant with a yield of 5, and exposure of this form
to ultraviolet light induced a further mutation, yielding 9.
Subsequent treatment with ultraviolet light and mustard gas
produced successive mutant strains with yields 20, 25, 30,
and eventually 50 times as much as that of the original strain.

The development of antibiotic-resistant strains of bacteria, of insects resistant to DDT and a host of other recently
discovered insecticides, and of rats resistant to poisons, are
genuine evolutionary changes. By indiscriminately spraying
insecticides, or by freely prescribing antibiotics and adding
them to animal feeds, we change the environment – and
natural selection ensures that if a species throws up advantageous mutations they are propagated and the genetic constitution of the species changes.

8.4 Shell pattern in snails

Cepaea nemoralis is a common snail in Britain and Europe,
widespread in woods, fields, hedgerows, scrubland and
sand dunes. The shells of this species are extraordinarily
diverse: the ground colour may be various shades of brown,
yellow or pink, and the shell may be plain or girdled by one
to five dark brown bands, of varying width (Fig. 8.7). One
population of snails will include a number of these variants,
and a neighbouring population may have a different set of
patterns. These shell patterns are inherited, and are
controlled by a complex of genes. The ground colour of
the shell is determined by one gene with at least seven
alternative forms (alleles). Dark brown is dominant to pale
brown and to three different pink alleles, and these are
dominant to two different yellows. In shell banding 'no
bands' is dominant to 'banded', two other genes control the
number of bands, four more genes control the width,
colour and intensity of the bands and another determines
whether each band is complete or interrupted.

Because the environments occupied by snail colonies
with different shell patterns seemed to be much the same
it was at first assumed that the variations were not
adaptive but random and due to genetic drift. More
thorough investigation has shown that much of this
variation is due to selection and that the more the situation

is investigated the more complicated it and the selective factors that may explain it become.

As in the peppered moth, birds are the selective agent responsible for some of the shell variation. Thrushes are fond of snails and in order to eat a snail a thrush will pick it up, carry it a short distance to a suitable stone and then drop or tap it on the stone until the shell shatters. The broken shells around these thrush 'anvils' provide a record of the snails collected and eaten by the birds, and the proportion of each variant can be compared with those in the surviving population. Comparisons of this sort, in a variety of habitats and at different times of year, show that thrushes pick out snails that contrast with the background – in woods with little undergrowth or on open downland banded shells are at a disadvantage, while unbanded shells are more conspicuous in woods with undergrowth and in scrub. The ground colour is also selected to match the habitat. But there is also a seasonal effect: in winter and early spring brown and pink shells are favoured; in late spring and summer, when the background is greener, yellow shells are favoured. In this way a polymorphic population may be maintained in one place because selection favours different forms at different seasons, or at different stages in the life cycle.

Bird predation is by no means the only factor influencing the snail polymorphism. Some genes for shell colour also have physiological effects; some variants have advantages in warm conditions and others in cold. In general, yellow shells are significantly more common in south-western Europe (Spain, southern France) than further north (Britain, northern Germany). Local variation within populations is influenced by microclimatic factors such as the accumulation of cold air in small hollows and the high midday surface temperatures of sand dunes in summer. A brown shell will absorb heat more efficiently than a yellow one, and so do better in the cold hollows and worse (through overheating) on the exposed dune tops. Another complicating factor is that birds may learn to recognize, and search for, the snail shell pattern that is most abundant; rare patterns may escape notice not because they are inconspicuous but because the bird does not recognize them as food. This sort of **frequency-dependent selection** will favour the rarer variants, but as they become more abundant their advantage will fall off.

There is also a large-scale polymorphism in *C. nemoralis*: many colonies over a large area show a certain uniformity and are separated from neighbouring population groups by quite sharp boundaries, where a different variation pattern takes over. The mechanism behind this is not fully understood, but fossil shells suggest that some of these 'area effects' have been stable for thousands of years and so are unlikely to be due to chance factors such as the expansion of small founder populations that happened to differ. Adjacent areas with different shell patterns may represent incipient geographical races.

We are still far from understanding or explaining all the variation in *Cepaea* shells, within and between populations – indeed, *Cepaea* polymorphism has been called 'a problem with too many solutions.' But the complexity of the interacting factors (predation, climatic and seasonal variation, multiple genetic control, physiological differences, frequency effects, etc.) makes *Cepaea* a more realistic example of natural selection than simpler situations such as sickle-cell anaemia or industrial melanism.

8.5 Social insects

Insects that live in highly organized societies – termites, bees, wasps and ants – have always fascinated naturalists. Darwin was the first to explain how the remarkable instincts governing insect societies could be produced by natural selection, and recently the sex ratio in social insects has been used as a test of a modern variant of selection theory – **kin selection**.

In the social Hymenoptera (the group of four-winged insects that includes ants, bees and wasps), each colony or nest contains one or more fertile females (the queens), a number of fertile males (the drones or kings), and many sterile females (the workers). In ants, the workers are often of several different kinds or castes – such as soldiers, indoor workers and outdoor workers – and these worker castes may differ in size and structure as well as in behaviour. The organization of insect societies may be very complicated, with strange echoes of traits of human societies. For example, some ants are stockfarmers, keeping aphids which they 'milk' for honeydew, care for in the nest and carry out to the food plants; others are gardeners, cultivating yeasts or other fungi in the nest. In some ant species the workers are incapable of caring for the nest or even of feeding themselves, but spend their working hours raiding the nests of other species and capturing the larvae or pupae, which they carry back to their own nest. When these captives hatch they behave as they would have done at home, caring for their captors and the captors' young, as slaves. In *The Origin of Species* Darwin wrote a fine account of these slave-making ants, and his observations of them. They attracted his interest because a system in which animals of one species devote themselves to the welfare of another seems inexplicable by natural selection, for selection can only promote the successful, and to live in slavery is no success.

Darwin was able to show that the dependence of slave-making ants on their slaves is less in one species than in another, and less in one geographic race of a species than in another. He guessed that slave-making originated in the habit of carrying home, as food, larvae and pupae of other species. Some of these might hatch and work, instinctively, as slaves, and natural selection would favour the slave-makers devoting themselves to the collection of pupae not as food but as potential slaves. The slaves themselves will be incapable of any long-term retaliation through natural selection, for they are sterile workers and cannot pass on their genes.

Yet the origin of these sterile worker castes in social insects poses another puzzle for natural selection: why should it favour the loss of reproductive potential in most individuals of a species? Darwin wrote of these sterile workers as a 'special difficulty, which at first appeared to me insuperable, and actually fatal to my whole theory.' The main difficulty he had in mind was not the sterility of the worker caste (we now know that this is usually the result of environmental factors, especially the food given to the larva), but the difference in form between the workers and their parents, and between the different castes of workers. For the structure of the workers could never be directly affected by natural selection because they produce no off-spring. Darwin's explanation included an analogy with the division of labour in human society, and the comment 'selection has been applied to the family, not the individual'. These ideas have been taken much further in the modern theory of kin selection.

In many ant species all the members of a colony share the same mother and father – the inseminated queen who founded the colony. Thus the colony is a family, consisting of brothers and sisters. This fact is the key to the altruistic behaviour of the workers, sterile females who devote themselves to the care of their mother and her offspring, their brothers and sisters. The colony can be thought of as one organism, in which the division of labour amongst the worker castes is analogous to the division of labour between different cell types or organs in an individual animal. Our blood cells and liver cells, for example, have lost the possibility of contributing to the next generation – except vicariously, through the egg or sperm cells that they may help to nourish. In the same way, worker ants cannot reproduce themselves, but they can help to contribute to the next generation by caring for the queen and her eggs and young, their brothers and sisters. The slave workers in the nests of slave-making ants are deceived by their instincts into caring for the young of another species: their altruism is misdirected. We can ask how close should the relationship be, between the altruist and the recipient of her kindness, for self-sacrifice to be worth while? The theory of kin selection answers this question.

In sexual species like our own, each individual receives half of his or her genes from one parent, and half from the other. Our genetic relationship to each of our parents, and to each of our children, is therefore $1/2$. Our parents, in turn, each received half of their genes from one of our four grandparents, so that our genetic relationship to each grandparent (and to each grandchild) is $1/4$. For brothers and sisters (excluding identical twins, where the relationship is one) the argument is a little more complicated because of the way in which the parental genes and chromosomes are shuffled in the production of eggs and sperm (see Section 5.4). Genetic relationship between brother and sister ranges from almost nil (in the very unlikely case when the two happen to receive opposite halves of both the maternal and paternal sets of genes) to almost one (when they receive the same half of both sets of genes – equally unlikely), but on average it is $1/2$, the same as the parent/child relationship. Your aunts and uncles, or nephews and nieces, share, on average, $1/4$ of your genes; the first-cousin relationship is $1/8$, the half-brother $1/4$, and so on. Given these degrees of relationship we can frame a theory of the evolution of altruism by natural selection. It is worth laying down your life to save three of your children, or three of your brothers and sisters, because their combined genes represent $1^{1}/_{2}$ of yours. It is not worth laying down your life for five cousins ($5/8$) or three grandchildren ($3/4$), although in the latter case you might reflect that if your reproductive life was finished and theirs just beginning the unselfish act would be to your genetic advantage.

Is there anything more to this theory than a morbid parlour game, or the basis for heartless jokes about mothers-in-law (relationship nil)? One situation has been found where the theory has consequences that can be tested by confrontation with nature. This test of kin selection is also a good example of a mathematical test of natural selection which is made in nature, and not on paper or in a laboratory experiment.

The test involves the social insects, and hinges on their peculiar method of sex determination. In Hymenoptera fertilized eggs develop into females and unfertilized eggs into males. This system is also found in some mites, and a few other insects. Why it developed is not known. It is called **haplo-diploidy**, for it means that females are diploid, formed by the fusion of two gametes, each carrying a haploid set of genes, and males are haploid, having only the genes from one gamete – the egg. Because of this, the genetic relationships in families of ants, bees and wasps are peculiar. Males, developed from unfertilized eggs, have received all their genes from the mother and have a genetic relationship with her of one, not $1/2$ as in other sexual organisms. Females receive half their genes from the mother and half from the father and so have the usual $1/2$ relationship with each parent but, because the father has only a haploid set of genes, all his sperm are identical and sisters are therefore related by $3/4$ (an average of $1/4$, through the mother as usual, added to $1/2$ through the identical paternal contribution), not $1/2$ as is usual. The sister/brother relationship is not the normal $1/2$, but $1/4$: the brother has only half as many genes as his sister and has received them all from the mother. They may be the same half as his sister has (relationship $1/2$; very unlikely) or the opposite half (relationship 0; equally unlikely), and the relationship averages out at $1/4$.

In an ant nest, therefore, the workers are caring for their mother (relationship $1/2$) and her offspring, their brothers (relationship $1/4$) and sisters (relationship $3/4$). In working to bring up sisters a worker can contribute more to the next generation than by investing an equal amount of energy in bringing up her own daughters (relationship $1/2$): hence the tendency for sterile female castes to develop in haplo-diploid insects.

The theory also says that workers should devote three times as much energy to raising fertile sisters (potential queens) as they should to raising males, because they share three-quarters of their genes with sisters and only a quarter with their brothers.

This prediction of the theory can be tested. A rough estimate of the energy invested by workers in raising fertile males and females is the weight of individuals of each sex produced, since the weight represents the result of feeding by the workers. Estimates of the sex ratio in nests of 21 species of ants have been collected and corrected by a factor allowing for the differences in weight between males and females in each species. The resulting figure, or 'invest-

ment ratio', averaged over all the species, was 1:3.45 – close to the prediction from theory of 1:3.

A most interesting control test is possible, by comparing this ratio with that in slave-making ants, where the workers are raising the offspring of another species, and so should have no interest in their sex ratio. In slave-making species the sex ratio should be controlled by the queen, who would benefit most from a 1:1 ratio since her relationship to both sons and daughters is $1/2$. The 'investment ratio' has been estimated in nests of two species of slave-making ant, and found to average 1:1. Even more remarkable is that one of these slave-making species enslaves workers of a species which produce a 1:3 ratio of investment when working for their own queen. These results are consistent with the theory of kin selection, and with no other theory yet conceived. How they are to be explained in detail is not yet known. For example, how can it be that potential slave ants 'know' that when they are in their own nest a 1:3 ratio of investment is to their advantage, and how can they be 'told', when enslaved, that their new queen 'wants' a 1:1 ratio? We can be sure that the words 'know', 'told' and 'wants' here are shorthand for some other expression, perhaps of the form 'natural selection has adjusted the genes governing behaviour, or governing the hormonal or nutritional regime of the workers'. In any case the social insects, far from being a 'special difficulty' (Darwin's words) for natural selection, are proving to be a gold mine for the theory, especially for the theory of kin selection. This theory, successfully applied to genetically controlled behaviour in insects, is by no means uncontroversial when applied to other animals. Its application to humans is mentioned in Section 14.2.

8.6   A cautionary tale

One of the classic cases of natural selection in the wild concerns the scarlet tiger moth, *Panaxia dominula*, in Britain. At Cothill, near Oxford there is a colony carrying a mutant gene, the *medionigra* gene, in which heterozygotes and homozygotes are easily recognized (Fig. 8.8). The colony has been studied by Oxford scientists for 50 years, both in the wild and in the laboratory. Several general conclusions have been reached.

One is that mutants have lower viability as eggs and/or larvae and lower fertility as adults than the wild-type, so that natural selection tends to eliminate the mutant gene, but the rarer mutant forms have an advantage at mating, because females seem to prefer males of a different phenotype. This rare mating advantage is an example of frequency-dependent selection and tends to maintain the mutant gene.

Another conclusion is that the frequency of the mutant gene in the Cothill colony increased from about 2% in the 1920s to about 10% in 1940, and then decreased again during the 1940s and 1950s. Similar declines were observed in colonies started elsewhere and in other colonies where the *medionigra* gene was introduced by releasing laboratory-bred stock. These declines in mutant frequency were assumed to be due to natural selection, as was the increase at Cothill during the 1930s, though the selective factor responsible for that increase has not been identified. All in all, work on the scarlet tiger moth has been seen as one of the best and most thorough investigations of natural selection in the wild.

An important part of the experimental work on the *medionigra* gene in the scarlet tiger moth has been the establishment of artificial colonies in suitable habitats, by releasing laboratory-bred larvae to establish populations

with a known initial frequency of the mutant gene. One such colony was started in 1961 on a stretch of disused railway line at West Kirby, near Liverpool, a part of the country where the moth is unknown in the wild. In 1961 about 13 000 caterpillars were released, with a gene frequency of *medionigra* of 25%. The colony was sampled in 1962–67 and the frequency of *medionigra* estimated at from 16% (1965) to 32% (1967). The colony was sampled once more, in 1976, when the frequency of *medionigra* had fallen to 9%. No further work was done on this experimental colony, and it was assumed to have become extinct. But in 1988 one *bimacula* (the homozygote) moth flew to a light trap at West Kirby and a search in 1989 showed that the colony still existed on the disused railway line. Thirty caterpillars were collected and reared yielding 28 adults (14 *dominula*, 12 *medionigra* and two *bimacula*), and 40 more adults were seen in the wild (22 *dominula*, 15 *medionigra* and three *bimacula*).

These samples give an estimate of *medionigra* gene frequency of 27%, virtually the same as the 25% in the original stock released 28 years earlier. This absence of change or stability over 27 generations (the scarlet tiger moth has one generation per year in Britain) is in complete contrast to the fluctuation over 20 years at Cothill and the decline in *medionigra* in other experimental colonies. The discoverers of the West Kirby scarlet tiger moth colony noted that there was one difference between that colony and the others studied around Oxford – lack of human interference, and by 'humans' they meant 'scientists'. The colony has been studied in the same way (by collecting and breeding caterpillars, and by observations in the wild) for seven more years (1990-96). The gene frequency of *medionigra* was estimated at between 15% (1990) and 26% (1995), with an average of 20%, from observations of 700 moths. Obviously, in these years, the colony was subject to human interference.

In physics, Heisenberg's uncertainty principle says (roughly) that we cannot gather data about something without interfering with it; that precision in measuring one aspect of a system is paid for by uncertainty about another aspect. Does a principle like this apply in evolutionary biology? Could it be that our hard-won 'knowledge' of natural selection on the *medionigra* gene in the scarlet tiger moth reflects interference by the scientist more than a natural process? The obvious answer is 'let's wait and see what happens to the West Kirby colony' – but how can we see what happens without interference?

Typical form

medionigra form

bimacula form

Figure 8.8

Scarlet tiger moth, *Panaxia dominula*.

Neutral evolution and the molecular clock

The two preceding chapters concentrated on natural selection, in theory and in the real world. There is no doubt that natural selection is a stringent monitor – or, in Darwin's words, 'that natural selection is daily and hourly scrutinising, throughout the world, the slightest variations; rejecting those that are bad, preserving and adding up all that are good.'

The picture of evolution by natural selection that comes out of classical genetics and the observation of populations is one of selection as a controlling force with two aspects, stability and change. Stability is maintained by stabilizing or negative selection, eliminating ill-adapted variants – 'rejecting those that are bad'. Change is promoted by directional or positive selection, favouring or 'preserving and adding up' variations better adapted or more efficient in the environments inhabited by their carriers.

From these two aspects of natural selection, we should expect the history of each species or lineage to be a mixture of episodes of change, when directional selection causes adaptive change – to exploit the environment more efficiently, to cope with changes in it or to permit invasion of new environments – and episodes of stability, when well adapted populations are maintained in stable environments. Those expectations relate to features that are under the scrutiny of natural selection, such as genetically controlled aspects of structure, physiology and behaviour. They began to be questioned in the 1960s, when the detailed structure of proteins from different species was first worked out.

9.1   Rates of protein evolution

Proteins – chains of amino acids – are a direct expression of the coded message in DNA, and in the 1960s the sequence

	Human	Elephant	Platypus	Ostrich	Starling	Crocodile	Lungfish	Coelacanth	Goldfish	Shark
Human		26	40	43	41	47	83	70	68	71
Elephant	18		45	45	48	50	84	72	63	74
Platypus	28	32		54	52	51	89	74	70	76
Ostrich	30	32	38		26	36	91	75	68	73
Starling	29	34	37	18		47	91	77	67	70
Crocodile	33	35	36	26	33		85	78	70	77
Lungfish	59	59	62	64	64	59		90	94	86
Coelacanth	49	51	52	53	54	55	63		83	78
Goldfish	48	44	49	48	47	49	66	58		88
Shark	50	52	54	52	50	55	63	55	62	

Table 9.1

Differences in the alpha haemoglobin chains of various vertebrates

The alpha chain contains 141 amino acids in most of these animals, but the lungfish has 143 and the coelacanth and goldfish have 142. The numbers above and to the right of the diagonal are the numbers of differences between the alpha chains, and the numbers below and to the left of the diagonal are the percentage differences.

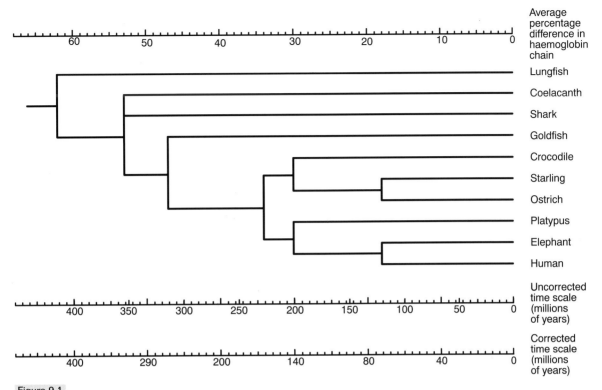

Figure 9.1

Family tree based on differences between the alpha haemoglobins in Table 9.1

The two time-scales are calibrated from the assumption (based on the earliest known fossils) that sharks, lungfishes and coelacanths diverged about 400 million years ago (60% difference in amino acids is set at 400 million years). The lower time-scale is corrected for 'multiple hits' (see text).

of amino acids in a protein was the closest we could get to reading that message. Some typical diferences are shown in Table 9.1. The unexpected finding is that the haemoglobin of a shark differs more from that of a goldfish or a lungfish than it does from that of a bird or a human being. We would expect that land vertebrates should have experienced more changes in their blood chemistry than have fishes because they have learned to breathe air — and birds and mammals have also become warm-blooded, with a much higher activity and rate of metabolism.

Comparing the different animals in Table 9.1 with their fossil relatives from Devonian times (more than 350 million years ago) sharks, coelacanths and lungfishes have hardly changed whereas birds and mammals have been totally transformed from fish-like ancestors. Yet the blood proteins have apparently changed just as much in fishes as they have in birds or mammals — the greatest difference in Table 9.1, 66%, is between goldfish and lungfish rather than between fishes and land vertebrates. Such observations led to the

idea that proteins such as haemoglobin evolve at a roughly constant rate in each lineage – the **molecular clock**.

In Table 9.1 the haemoglobin of human and elephant differs by 18%, as does that of the ostrich and starling. This implies that the human and elephant lineages diverged from an ancestral mammal at about the same time as the ostrich and starling lineages diverged from an ancestral bird. The platypus differs from the elephant and human by about 30%, these three mammals differ from the birds by an average of 33%, and the crocodile differs from the birds by about 30% and from the mammals by an average of 35%.

We could use these figures and those for the four fishes to make a rough family tree (Fig. 9.1). Fossils show that sharks, coelacanths and lungfishes were distinct 400 million years ago, so we might infer that the platypus separated from the other mammals and the crocodile split from the birds about 200 million years ago, and that elephants and humans separated about 120 million years ago, at the same time as starlings and ostriches split. But these estimates would be inaccurate even if the clock runs true. When two sequences differ by as much as 60% (like the lungfish and shark haemoglobins) we must assume that the real number of differences between them is greater because many of the observed differences will be at sites that have suffered 'multiple hits' – more than one change. An analogy would be to draw 60 tickets from a drum full of tickets numbered from 1 to 100. In the sample some numbers would occur twice or more, just by chance. We can correct for this source of error, and in this case it gives divergence times of about 140 million years for the platypus/mammal and crocodile/bird splits, and about 70 million years for the elephant/human and ostrich/starling splits.

How could the rate of haemoglobin evolution in each lineage be constant? The classical explanation would be Darwin's – natural selection – with the implication that the differences between the variant haemoglobins are adaptive, fitting each species for its particular way of life. And a theory that will explain clock-like evolution by natural selection has been proposed. It is named the 'Red Queen', after the character in *Alice's Adventures in Wonderland* who said 'Now here, you see, it takes all the running you can do, to keep in the same place.' The idea is that if one species or lineage achieves some successful adaptive change, and so becomes fitter in the struggle for existence, its competitors – for food or other resources – will suffer a decrease in fitness, and if they are to survive must make some compensating change. We can imagine the whole of life as linked through a web of competition, so that animals as physically remote as a shark and an ostrich are connected through many intermediates in the struggle for resources. On this view every species must be doing 'all the running' it can, just 'to keep in the same place' if it is to survive, and the result might be a roughly constant rate of change, in haemoglobin or any other feature.

9.2 Rates of DNA evolution

The alternative explanation for a constant rate of evolution in proteins is much simpler. It is that most of the differences between the variant haemoglobins are not adaptive but neutral (or nearly so) and are therefore not distinguishable to natural selection. We saw in Section 7.2 that neutral mutations will occasionally become fixed or universal in a lineage simply by chance (genetic drift), and that the average number of generations between the appearance of a neutral mutation and its fixation is four times the population size. In one of the simplest and most remarkable bits of mathematical genetics, the Japanese scientist Motoo Kimura showed that the expected rate of fixation of neutral mutations is equal to the rate at which such mutations occur, independent of population size. **Mutation rate** here means the rate per gamete (sperm or egg), and is a very small number (page 32) because mutations are rare events. But, as far as we know, the mutation rate is constant and, if it is, the equality between it and the rate of fixation of neutral mutations implies a constant rate of change, a molecular clock of evolution. Ideas like these led Motoo Kimura to propose the neutral theory of molecular evolution in the late 1960s. The provocative form of his theory can be summed up in one statement: 'molecular changes that are less likely to be subject to natural selection occur more rapidly in evolution.'

There are two issues here:

- whether a molecular clock exists – whether genes, or bits of genes, or lengths of DNA evolve at a roughly constant rate in different lineages;
- if a clock or an approximate clock exists, whether it is to be explained by natural selection or by neutral change – whether the changes in DNA are adaptive or non-adaptive.

Answers to these questions began to appear through a series of technical advances and discoveries in the late 1970s and 1980s. The first of these advances was the invention of ways of sequencing DNA so that we can read the genetic message directly rather than at second hand by sequencing proteins. Methods of sequencing DNA came into general use at the end of the 1970s, and rapidly led to a host of discoveries about the structure and organization of the genome. Two of these discoveries are particularly important here.

The first is the generality of 'genes-in-pieces,' the fact that in virtually all genes the coding sequence specifying the protein is not a continuous stretch of DNA but is broken up, with coding sections (exons) interrupted by non-coding sections (introns) (Fig. 4.11). The coding sections are typically 40–50 amino acids long (120–150 base pairs; three bases to each amino acid) but the intervening introns range from about 50 bases to several thousand. Thus a typical gene, such as that coding for alpha haemoglobin (which is about 150 amino acids long) is actually ten or more times longer than the expected 450 or so bases (three times the number of amino acids). The globin genes of vertebrates each consist of three exons (coding sequences) interrupted by two non-coding introns, and the positions of the introns are the same in all globin genes.

The second important discovery to come from DNA sequencing is the existence of **pseudogenes**. A pseudogene is a 'silent gene' or a 'dead gene' – a stretch of DNA resembling a functional gene but having some defect that makes it non-translatable or useless. Pseudogenes are of two types, classical and processed. A **classical pseudogene** resembles a functional gene nearby on the chromosome but is silenced by defects such as a mutation in the site that initiates transcription, a mutation that produces a 'stop' codon early in the sequence, or a single-base deletion that changes the reading frame (Box 6.1). Most pseudogenes show several of these defects. A **processed pseudogene** may be on a different chromosome from the functional gene that it resembles and differs from it first by having only the coding portions – it lacks the introns and the starting sequence containing the initiation and promoter signals – and second by having a 'tail' sequence that is missing in the functional gene. This 'tail' and the lack of introns show how processed pseudogenes originate because during transcription (the process in which the DNA message is transcribed into messenger RNA) the introns are spliced out and a tail is added. A processed pseudogene therefore represents a transcribed copy of a functional gene which, through some accident, has been copied back into the DNA.

'Genes in pieces' and pseudogenes are important in the study of molecular evolution because they allow tests of the relative importance of natural selection and neutral drift at the DNA level, rather than in features of the organism where the two are hard to differentiate (Section 7.2). If the DNA sequence of a gene is known in two or more species the relative rates of change of exons and introns can be estimated from the number of differences between them in the different species. In the same way, if the DNA sequence of a pseudogene in two or more species is known its rate of change can be compared with that of the related functional gene. A third comparison that may be useful here is to compare the rate of change at the first, second and third positions in the coding triplets of exons. As described on page 20, the genetic code is redundant, with most amino acids being coded by more than one triplet. This redundancy means that there are two kinds of point mutation: those that alter the amino acid coded by a triplet and those that do not. The first kind replaces one amino acid by another, the second is silent or synonymous, causing no change in the protein. Most silent mutations involve the third position of the triplet. Figure 4.9 shows that five of the 20 amino acids (valine, proline, threonine, alanine, glycine) are each coded

by a block of four triplets, so that the first two nucleotides are enough to specify the amino acid and any third position mutation will be silent – the third position is **fourfold degenerate**. Isoleucine is coded by three triplets and will work with three different nucleotides in the third position (it is **threefold degenerate**). Nine amino acids (phenylalanine, tyrosine, histidine, glutamine, asparagine, lysine, aspartic acid, glutamic acid, cysteine) are coded by two triplets, permitting two nucleotides in the third position, which are therefore **twofold degenerate**. Two amino acids (methionine, tryptophan) are coded by only one triplet and do not permit silent mutations. The three remaining amino acids (leucine, serine, arginine) have six different codons and so not only have four codons in the fourfold degenerate third positions but also may allow a change in the first position. If all triplets and all mutations were equally probable (except the 'stop' triplets and mutations producing them, because a 'stop' will cut the protein short and is usually lethal), we should expect about 70% of third position mutations and 5% of first position mutations to be silent, whereas all second position mutations are non-degenerate and alter the amino acid.

A first test of the neutral theory is to consider differences in rate of change in fourfold degenerate, twofold degenerate and non-degenerate (replacement) positions. For example, about 40 different genes from three groups of mammals – humans, rodents (e.g. mouse, hamster) and artiodactyls (e.g. pigs, goats, cattle) – have been compared. Assuming that these three groups diverged 80 million years ago and calling the non-degenerate rate 1.0, the average rate over the 40 genes for twofold degenerate sites is 2.4 and that for fourfold degenerate sites is 4.5. If the average non-synonymous or replacement rate is 1.0, the average synonymous or silent rate will be 5.3. So here we estimate that twofold degenerate sites evolve more than twice as fast as non-degenerate ones, fourfold degenerate sites evolve more than four times as fast, and the total silent rate is more than five times the replacement rate. These results seem inexplicable by natural selection but match the prediction of neutral theory that 'molecular changes that are less likely to be subject to natural selection occur more rapidly in evolution.'

Next we can compare rates of change in coding and non-coding regions of genes. One study tested the haemoglobin genes from various groups of placental mammals against those of marsupials. Calling the non-degenerate or replacement rate 1.0, the synonymous rate in coding regions was 5.6 and the

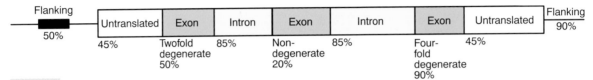

Figure 9.2

Rates of evolution in different parts of a gene

A typical mammalian gene (compare Fig. 4.11). The rates of evolution of the various regions are expressed as percentages of the rate of evolution of pseudogenes in mammals. In the three exons (parts of the gene that code the amino acids) rates are shown for three kinds of bases (see text) that occur within each exon, and are not separated as implied in the diagram. Fourfold degenerate sites are all in the third position of codons, twofold degenerate sites are in the first and (mostly) third positions, and non-degenerate sites are in the first and second positions.

rate in non-coding regions 5.3, so the non-coding regions evolve almost as fast as the silent sites in coding regions.

Finally, pseudogenes appear to evolve faster than any part of functional genes. If we call the pseudogene rate 100, we can estimate the percentage of that rate in various parts of an average mammalian gene (Fig. 9.2). The rates are fastest in the flanking sequence at the far end of the gene and the fourfold degenerate sites, both at about 90%. Introns come next, at 85%, and then the flanking sequence before the beginning of transcription (containing the promoter region) and twofold degenerate positions, both at about 50%. The transcribed but untranslated regions before and after the coding sequences are slower, at about 45%, and the slowest rate of all (20%) is for non-degenerate (replacement) positions.

9.3   Neutral evolutionary change

The explanation for this flood of numbers is satisfyingly simple. The fastest rates of change in recognizable genes or parts of genes occur in pseudogenes and, as far as we know, pseudogenes are without function because they have defects that prevent transcription or translation. They are therefore immune to natural selection because they are not expressed in the phenotype, and must evolve by random drift, since mutations in them will be neutral. The rate of evolution in pseudogenes will be controlled by the mutation rate. We know that some DNA evolves even faster than pseudogenes, like the minisatellite DNA which is used in genetic fingerprinting (page 27) where there is a very high mutation rate, up to 10% per gamete. If pseudogenes show the maximum neutral rate of evolution under normal mutation rates, parts of genes that change more slowly than pseudogenes must do so because they have some function and are therefore constrained by stabilizing selection. The difference between the rate of change of pseudogenes and

the different parts of functional genes is thus a measure of functional importance. The most important parts of a gene are those least subject to change — the non-degenerate positions in coding sequences — because any change in them alters the amino acid in the coded protein. The least important parts of a gene are those most subject to change — the flanking sequence at the end of the gene and the fourfold degenerate or silent positions in coding sequences. Because those silent sites change a little slower than pseudogenes (at about 90% of the pseudogene rate in mammals) they must be under some constraint, presumably selection caused by the availability of different transfer RNAs (page 20).

Here we have an explanation, at the DNA level, for how a molecular clock might run. Imagine the stretch of DNA coding the alpha haemoglobin gene in the common ancestor of two of the animals in Fig. 9.1, say the elephant and human. The clock 'ticks' every time one of the two descendent lineages substitutes a different base somewhere in that stretch of DNA. In Table 9.1, in which we are comparing only the amino acid sequences coded by that DNA, we find that they differ at 26 positions, or 18% of the amino acids in the protein. These differences reflect substitutions at non-synonymous (non-degenerate or replacement) sites in the coding DNA, the most slowly evolving bits of the gene. In mammalian alpha haemoglobins the average synonymous or silent rate of change in the coding regions is about seven times the non-synonymous rate, and the rate in the non-coding regions is almost as fast. As the non-coding regions are much longer than the coding regions the differences between elephant and human alpha haemoglobins shown in Fig. 9.1 will represent only a small fraction of the differences between the two genes. Most of those differences are unlikely to have been subject to selection because they are not reflected in the phenotype — the protein chain — and so will have been 'ticked up' in clock-like fashion.

Just as differences in rate of change between the parts of a gene are a rough measure of their functional importance, or the selective constraints on them, differences in rate of change between different genes are a rough measure of their importance. Estimates of rates of non-synonymous and silent change for some mammalian genes are shown in Table 9.2.

The most stable gene in Table 9.2 is that coding a histone, a protein that binds to DNA and helps to give the chromosome its organization; change in its structure seems virtually forbidden by selection. The least stable gene is that coding an interferon, an antiviral protein – here one might guess either that the protein is weakly constrained by selection or that, as part of the organism's defences, selection causes rapid change in the protein to meet rapid change in viruses. Notice that the hormone insulin appears twice in Table 9.2, both as the second most slowly and the second most quickly changing gene. Active insulin, which is produced by the pancreas and regulates blood sugars, is a rather short protein, 51 amino acids long, but the insulin gene generates a longer protein, proinsulin, from which a central portion (the signal protein) is spliced out to give the active hormone. This discarded, apparently functionless part of the protein evolves about seven times as fast as the functional portion. Notice also that the replacement rate varies a hundredfold (from less than 5% to 500% of the alpha haemoglobin rate) but the silent rate varies only fivefold (from 44% to 220%). Finally, the table shows that there is no universal DNA clock; each gene has its own clock, just as the different parts of a gene run at different rates (Fig. 9.2) or by different clocks.

From all of these comparisons a picture of molecular evolution emerges that is very different from the world of peppered moths and sickle-cell haemoglobin. In general, it seems to be true that *molecular changes that are less likely to be subject to natural selection occur more rapidly*. A stretch of DNA that has no function, such as a pseudogene, will change over many generations solely by the chance fixation of mutations that are invisible to natural selection. This is change by mutation pressure rather than by selection pressure. The 'Red Queen' says that 'it takes all the running you can do just to stay in the same place' but studies of DNA say something different – that you can stay in the same place only by tight functional constraint, and if that is lacking DNA will do 'all the running it can' purely by chance.

This is not to deny that positive or directional selection operates at the molecular level. A few examples are known in the haemoglobins, the most thoroughly studied molecules, mostly concerning changes in one or two amino acids which increase the affinity for oxygen in the haemoglobin of animals experiencing high altitudes such as llamas in the Andes or the Bar-headed Goose, which migrates across the Himalayas at heights of up to 9000 m. But those examples are not numerous. Most of the differences between haemoglobins of different species are likely to be neutral; for example among the alpha haemoglobins of the animals in Table 9.1 there are variant amino acids in over 80% of the sites in the molecule, and it is difficult to believe that all those variations are adaptive. In much the same way, the 550 different mutations known in human haemoglobin (page 30) seem to be either neutral or deleterious.

All this implies that a molecular evolutionary clock runs in the same way as an ordinary clock – by 'running down' (the clock on the mantelshelf won't work unless you wind it or give it a new battery, and even an atomic clock runs by radioactive decay). So, what powers molecular evolution?

Gene product	Number of triplets in gene	Replacement rate relative to alpha haemoglobin	Silent rate relative to alpha haemoglobin
Histone H4	101	0.048	1.6
Insulin	51	0.29	1.4
Parathyroid hormone	90	0.79	0.44
Growth hormone	189	1.7	1.1
Alpha haemoglobin	141	1.0	1.0
Beta haemoglobin	144	1.6	0.75
Albumin	590	1.6	1.7
Immunoglobulin V	100	1.9	1.4
Insulin signal protein	23	2.1	0.62
Interferon gamma	136	5.0	2.2

Table 9.2

Estimated rates of change of some mammalian genes

The lower the rate, the more stable, and therefore important, the gene.

10 Gene families and gene evolution

Chapter 9 introduced some of the ideas about evolution that have emerged during the last 10–20 years as the structure of DNA has become better understood. There is an obvious mismatch between the world of organisms – peppered moths, snails, social insects, etc. – and the world of DNA. Natural selection among organisms seems to be a stringent monitor, 'daily and hourly scrutinising' each variant (in Darwin's words). But variants in DNA may be invisible to selection and many are propagated and persist purely by chance.

As one example, there is a pseudogene (a dead or inactive gene; page 69) that resembles the gene for beta haemoglobin and is shared by humans, apes, old and New World monkeys and lemurs. With this distribution the pseudogene must have been in existence for over 50 million years (fossil apes were quite diverse 22 million years ago; fossil lemurs are not known, but lemurs are confined to Madagascar, which has been separate from Africa for at least 50 million years). How and why should an apparently functionless gene become fixed in a lineage, and persist for such enormous periods of time? To explain this it is necessary to introduce some fairly complicated ideas, but they will help to illustrate some of the important features of gene evolution.

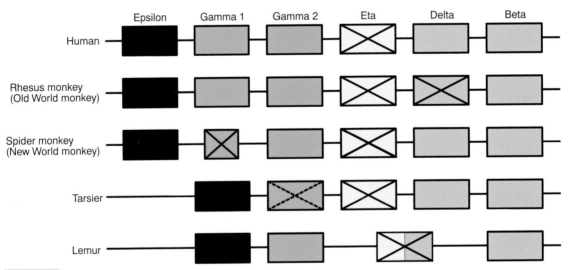

Figure 10.1

Maps of the beta haemoglobin gene family in some primates

Pseudogenes (silenced by mutations that prevent translation) are shown by crosses; the dotted cross on the tarsier gamma gene means that it is probably silenced, but this is not yet confirmed. The maps are aligned on the eta pseudogene and show the relations rather than the exact spacing of the genes.

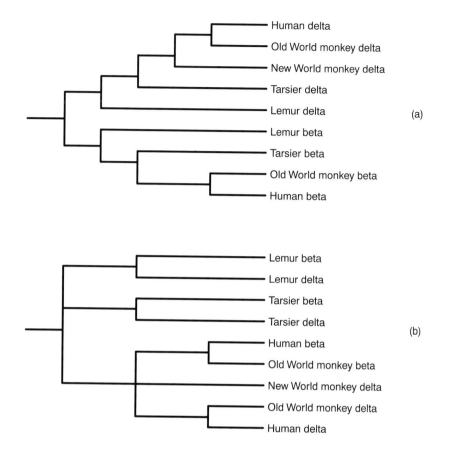

(a)

(b)

Figure 10.2

Gene trees and species trees

Trees based on DNA sequences of the non-coding (a) and coding (b) parts of the delta and beta haemoglobin genes of the animals in Fig. 10.1. The New World monkey beta gene is missing from the trees because its sequence is not yet known.

10.1 The beta haemoglobin gene family

Figure 10.1 shows the surroundings of the pseudogene in various beta haemoglobins. Like people, most genes exist in families and (as with people) there are different degrees of relationship between members of a family, the relationships being a consequence of history. So we can draw a family tree of genes, just as we can for a human family. The human beta haemoglobin gene family occupies about 60 000 bases of DNA and contains six genes, one of which is the pseudogene. Less than 10% of the bases code for proteins, and the five functional genes are arranged along the chromosome in the order in which they are expressed in life. The first gene, epsilon, is expressed early in embryonic life. Epsilon haemoglobin has a higher affinity for oxygen than adult haemoglobin, and so the embryo can extract oxygen from the mother's (adult) blood. The two almost identical genes for gamma haemoglobin are expressed in the fetus, later in prenatal life. Then comes the pseudogene named eta. The

two genes that are expressed in the adult, delta and beta, follow. The beta gene provides our major haemoglobin, and delta haemoglobin is made in only 1% of the quantity of beta. Human delta and beta haemoglobins differ by ten amino acids but they are functionally indistinguishable – if they were made in equal quantities, sickle cell anaemia (page 48), a mutant beta, would be no disadvantage, for delta would compensate for it.

In the Old World monkeys (baboons, macaques, rhesus monkey etc.) the organization of the beta haemoglobin family is the same as in humans, but the delta gene is silenced (it is a pseudogene). In New World monkeys (spider monkeys, marmosets, etc.) there are two gamma genes, as in humans and Old World monkeys, but one is silenced. The delta gene is also more active, producing about 6% of the adult haemoglobin. In the tarsier there is only one gamma gene, and it may be a pseudogene because there are mutations in the promoter region that

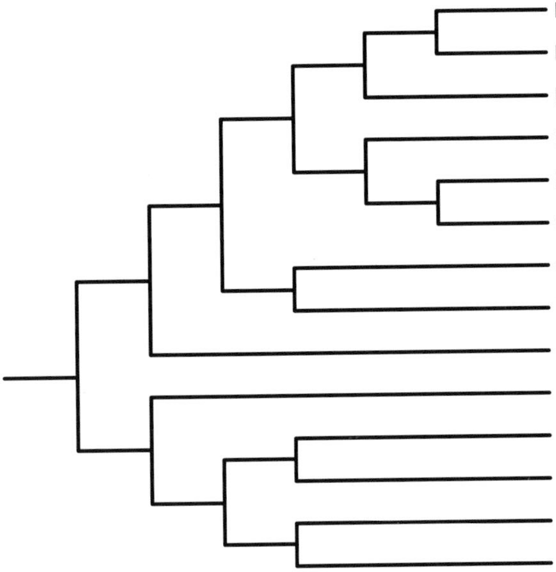
Human gamma 1
Human gamma 2
Mouse gamma
Mouse epsilon
Human epsilon
Goat epsilon
Goat eta
Human eta pseudogene
Opossum epsilon
Opossum beta
Mouse beta
Human beta
Human delta
Rabbit delta

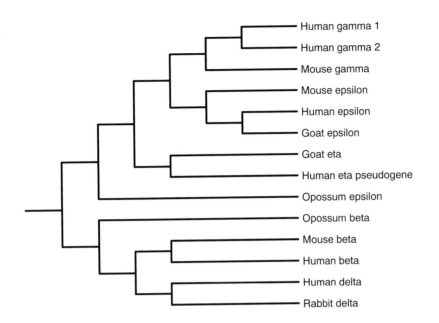

Figure 10.3

Gene relationships

A family tree based on DNA sequences of some of the genes in the beta haemoglobin family of mammals (compare Fig. 10.2).

should prevent or impair its expression. In tarsiers the delta gene is more active, producing about 20% of the adult haemoglobin. Finally, the lemur has only four genes in the family, and the one between gamma and beta, which occupies the position of both the pseudogene and delta in the other animals, is inactive. In other mammals, such as rabbits, rodents and goats, there are generally five genes in the family, with the first three (gamma, epsilon and eta) expressed in the embryo and the last two (delta and beta) expressed in the adult.

Figure 10.2 shows family trees for the non-coding and coding parts of the delta and beta genes of the animals in Fig. 10.1. The non-coding tree matches our ideas about how these animals are related, with the beta and delta genes each showing the same pattern. In the non-coding tree the tarsier beta gene (for example) is related not to its neighbour on the chromosome (the tarsier delta) but to other beta genes; the human and Old World monkey beta genes are related not to the delta genes of humans and monkeys, but to other beta genes. It seems that the non-coding parts of the genes have diverged over time, and show the same pattern of relationships or family tree as do other genes, and as do features of the bones and teeth. But the coding parts of the beta and delta genes have remained similar to each other in the lemur lineage, in the tarsier lineage and in the lineage containing humans and monkeys.

10.2 Gene duplication and gene conversion

To explain these observations on haemoglobin genes we need two mechanisms: **gene duplication** and **gene conversion**. Gene duplication (page 32) occurs through unequal crossing-over, when one of the two daughter chromosomes in a cell division receives two copies of a gene and the other receives none. The six genes in the human beta haemoglobin family are the duplicated descendants of one original beta haemoglobin gene. Their probable history can be inferred from the family tree of genes in Fig. 10.3. Marsupials (the opossum in this figure) have only two genes in the beta family, an embryonic (epsilon) and an adult (beta) gene.

The marsupial beta gene is related to both the beta and delta genes of humans, rabbits and mice, so the beta and delta genes must have arisen by a duplication after placental mammals separated from marsupials. Similarly, the marsupial epsilon gene is related to the gamma, epsilon and eta genes of placentals, so those three genes must have arisen by two successive duplications after placentals and marsupials separated. One other duplication must have occurred to produce the two gamma genes seen in humans, apes and monkeys. The human pseudogene is most closely related to the eta gene in goats, a functional gene which is expressed in the fetus, so the pseudogene in primates

(Fig. 10.1) evidently arose by silencing of a gene formerly functional in fetal life.

Gene duplication will explain the array of beta family genes in primates (Fig. 10.1), but we also have to explain why the coding parts of the beta and delta genes show a different pattern of relationships from the non-coding parts (Fig. 10.2). The problem here is that non-coding parts reflect the history of the species but the coding parts imply a different history, with the beta and delta from one animal (lemur or tarsier, for example) related to one another rather than to other beta and delta genes. This can be explained by gene conversion (also known as **concerted evolution**). The mechanism behind gene conversion is not yet fully understood, but it results in the transfer of information from one segment of DNA to a nearby segment, as if one were 'correcting' the other, so that the two evolve in concert rather than diverge over time. The symptom of gene conversion is that one part of a gene shows much greater similarity to the corresponding part of a neighbouring gene than would be expected from the history of the gene as a whole. Gene conversion has been very common in the beta haemoglobin family of primates: it has occurred repeatedly between the two gamma genes in the human lineage and in the Old and New World monkey lines, and between the delta and beta genes in the lemur lineage, the tarsier lineage and the line leading to humans, apes and

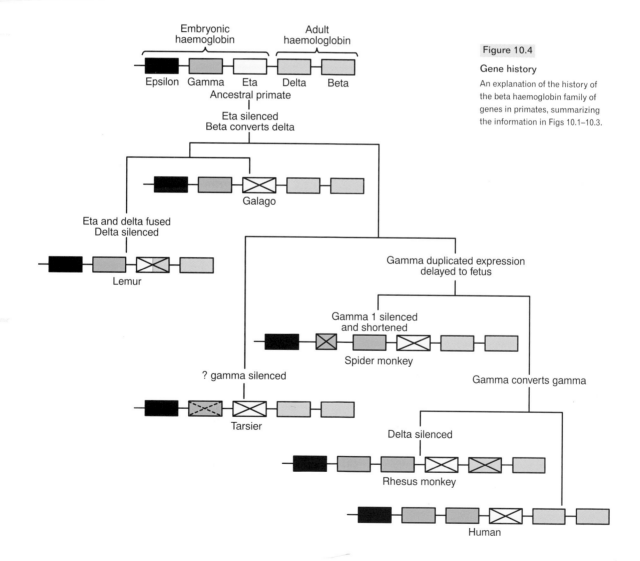

Figure 10.4

Gene history

An explanation of the history of the beta haemoglobin family of genes in primates, summarizing the information in Figs 10.1–10.3.

monkeys. But it seems, although we don't know why, that gene conversion is possible only between very similar sequences. After a gene duplication the duplicates will gradually diverge, by drift alone or by selection for different functions, *unless* conversion between them 'corrects' the divergence.

As time passes the least important (non-coding) parts of the two duplicates will diverge so far that conversion between them becomes impossible and only the coding and control regions, tightly constrained by selection, will remain similar enough for conversion to be possible. This is why the non-coding parts of the beta haemoglobin family (Fig. 10.2(a)) show the history of the species whereas the coding parts (Fig. 10.2(b)) have stayed similar enough for conversion to be possible, and so reflect the concerted evolution of the delta and beta genes rather than their independent history.

Now we can begin to answer the question that got us into these byways of gene duplication and gene conversion: why has an apparently functionless pseudogene persisted for some 50 million years in primates, from lemurs to humans? Figure 10.4 is a version of Fig. 10.1, with added information about the expression of genes (fetal or adult) and conversions between them. In all the animals except the lemur the pseudogene separates the genes expressed in the embryo (epsilon) or fetus (gamma) from those expressed in the adult (delta and beta), and all the events of gene conversion have been between the pairs of genes on either side of the pseudogene (the two gammas upstream from it and the delta and beta downstream from it). When something survives for as long as the pseudogene in primates, one would guess that it must serve a purpose. Our best guess in this case is that the pseudogene is useless in itself but is useful as a spacer, maintaining separation between the block of two or three functional fetal genes and the two adult genes. Its use as a spacer might be to prevent or inhibit gene conversion between those two functional blocks.

The lemur is the exception to these ideas because it has one fewer gene than the other animals. Figure 10.4 also shows the beta haemoglobin family in the galago, a relative of lemurs, in which we find the same five genes as in the New World monkeys, showing that the lemur is indeed the exception. The reason for this is found by comparing the sequence of the lemur genes with those of the other primates: the first one-third of the third gene in the lemur family matches the first one-third of the pseudogene in

other primates and the remaining two-thirds of the gene match that part of the delta gene in other primates. The lemur gene is thus a hybrid, caused by deletion of the stretch of DNA between the middle of the pseudogene and the middle of the delta gene. Because there are defects in the control signals in the pseudogene part of this hybrid, it is not translatable and remains a pseudogene. Hybrid haemoglobin genes of this sort, caused by unequal crossing-over, are well known in humans. Eight different human chromosome mutations are recorded in which there is a hybrid delta/beta or beta/delta gene — three produce a haemoglobin with the first half delta and the second half beta, four produce one with the first half beta and the second half delta, one is mostly delta with a chunk of beta in the middle. Figure 10.5 shows how unequal crossing-over between the delta and beta genes would produce the two main types of hybrid genes in humans, and how a similar event will explain the hybrid pseudo/delta gene in lemurs: an accident in crossing-over in an ancestral lemur resulted in both descendent chromosomes containing a fusion between the pseudogene and the delta gene. In one of the descendent chromosomes the delta gene is inactivated (by loss of its first one-third). Although this is apparently a harmful mutation it is the pattern that has proliferated and persisted in lemurs. Duplication of the gamma gene in monkeys, apes and humans may be explained by the same sort of accident (see Fig. 10.7).

10.3 More gene families

The beta haemoglobin gene family in humans and other mammals can be understood as the result of successive duplications, conversions, and silencing (producing a pseudogene) over a long history beginning with the two genes now found in marsupials.

These features are quite typical. The closest relative of the beta haemoglobin family is the alpha haemoglobin family, found on a different chromosome in mammals. In humans it contains two identical alpha genes, which produce the adult alpha haemoglobin, two alpha pseudogenes, a gene named zeta that produces an embryonic haemoglobin, a zeta pseudogene and a functional gene named theta. There is a story of duplication, conversion (to maintain the identity between the two alpha genes) and silencing behind this gene family like that behind the beta family. In birds, as in mammals, the alpha and beta haemoglobin gene families are on different chromosomes, but in frogs the alpha and beta haemoglobin genes are closely linked on the same chromosome. This implies that the alpha and beta families are descendants of what was once a single family. The closest relative of the haemoglobins is myoglobin (Figs 4.5 and 4.6), found in muscle, which is coded by a gene with exactly the same organization as alpha and beta haemoglobin, with two introns and three exons. Whereas

alpha and beta haemoglobin are found only in fishes and land vertebrates, haemoglobin-like genes and proteins occur in insects and other invertebrates – and even in plants, where they are best known in the root nodules of legumes (clovers, beans, etc.), structures that permit these plants to fix nitrogen. It was thought at one time that the

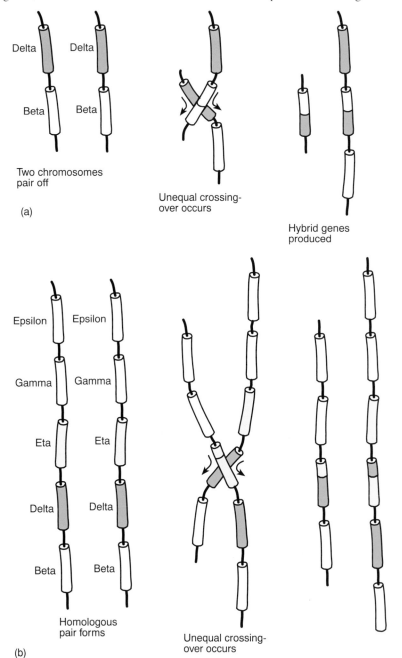

Figure 10.5

Chromosome mutations in haemoglobin genes

Unequal crossing-over could explain how the hybrid beta/delta haemoglobins in humans (a) and the pseudo/delta haemoglobin gene of lemurs (b) arose.

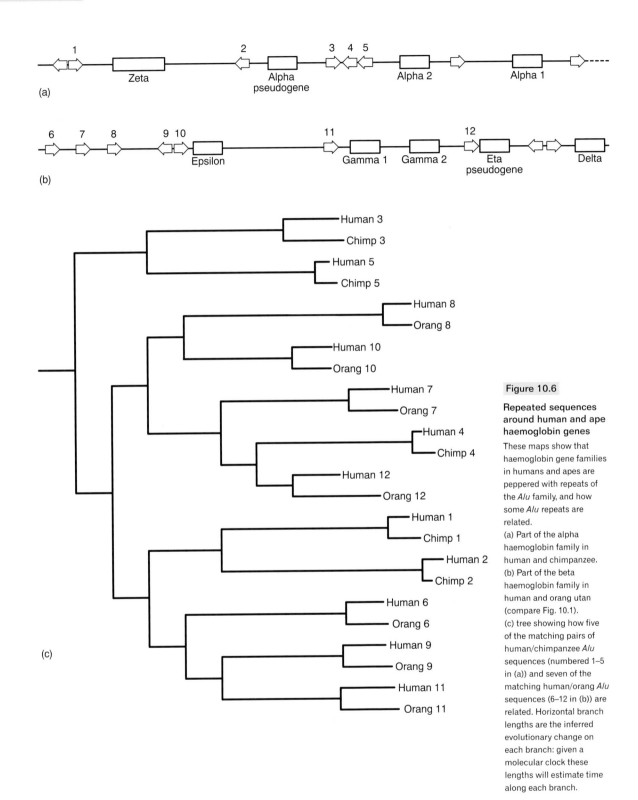

Figure 10.6

Repeated sequences around human and ape haemoglobin genes

These maps show that haemoglobin gene families in humans and apes are peppered with repeats of the *Alu* family, and how some *Alu* repeats are related.
(a) Part of the alpha haemoglobin family in human and chimpanzee.
(b) Part of the beta haemoglobin family in human and orang utan (compare Fig. 10.1).
(c) tree showing how five of the matching pairs of human/chimpanzee *Alu* sequences (numbered 1–5 in (a)) and seven of the matching human/orang *Alu* sequences (6–12 in (b)) are related. Horizontal branch lengths are the inferred evolutionary change on each branch: given a molecular clock these lengths will estimate time along each branch.

plant haemoglobins might have entered these plants quite recently, by **transfection** or gene transfer from an animal, perhaps a sap-feeding insect. But plant haemoglobins turn out to differ from those of animals in having one more intron, at a position where animal haemoglobins were suspected to have lost one. So plant and animal globins are probably related by descent, and in plants we find globin gene families just like those in animals. For example, in soya bean there are two clusters of globin genes, one containing four genes (the third of which is a pseudogene) and one containing two (the second a pseudogene), and in two other places on the chromosomes there are truncated (and so non-functional) fragments of globin genes. Evidently the human beta haemoglobin family (Fig. 10.3) is merely one small twig on a family tree of genes that ramifies through the animal and plant kingdoms, with repeated duplications causing branching of the same tree.

For a gene family to have as a long a history as the haemoglobins, it must have a function that is both promoted and conserved by natural selection. Not all gene families are like that. Scattered through the human beta haemoglobin gene family are a dozen short sequences, each about

300 bases long, belonging to the *Alu* family. In the human alpha haemoglobin family there are eight *Alu* sequences. All these *Alu* sequences are similar but they differ in detail by about 20%. As shown in Fig. 10.6, exactly the same *Alu* repeats are found in the beta gene family of orang utan and the alpha gene family of chimpanzee. In orang utan and chimpanzee, just as in humans, some of the *Alu* sequences run upstream (right to left in Fig. 10.6) and some run downstream. Members of the *Alu* family are dispersed throughout human DNA and account for almost 5% of it – there are about a million copies. The matching pairs of human and chimpanzee *Alu* sequences differ, on average, by about 2%, and the human/orang pairs differ by about 4% (their divergence is shown by the length of the lines linking the pairs in the tree in Fig. 10.6). These are the observations: how can they be explained?

The highly repetitive *Alu* sequences in human DNA cannot all be explained by gene duplication, because the usual mechanism of duplication (Fig. 10.5) will produce only repeats that face in the same direction, whereas six of the 17 *Alu* repeats in Fig. 10.6 face one way and 11 face the other. Instead, we must turn to a mechanism mentioned on page 21 that generates processed pseudogenes: after a gene has been transcribed into messenger RNA, with its introns

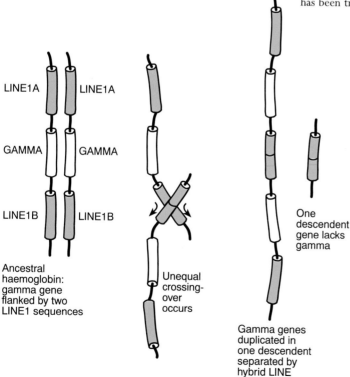

LINE1A   LINE1A

GAMMA   GAMMA

LINE1B   LINE1B

Ancestral
haemoglobin:
gamma gene
flanked by two
LINE1 sequences

Unequal
crossing-
over
occurs

Gamma genes
duplicated in
one descendent
separated by
hybrid LINE

One
descendent
gene lacks
gamma

Figure 10.7

Repeated sequences and gene duplication

How the gamma haemoglobin gene was duplicated in an ancestor of monkeys, apes and humans.

spliced out and a 'tail' added, that RNA is copied or inserted into a chromosome. The *Alu* sequences show the 'tail' typical of processed pseudogenes, and are scattered throughout human chromosomes. So the human *Alu* family seems to consist of about a million processed pseudogenes. The source gene (or genes) has not yet been identified, and we have no idea why it (or they) should generate so many pseudogenes. But that source is apparently important and strongly conserved because humans and Old World monkeys, which have been separate lineages for over 20 million years, have many virtually identical *Alu* sequences, which can only be recently inserted copies from source genes that have remained identical in monkeys and humans.

Once an *Alu* sequence is inserted in a chromosome it may be duplicated or deleted by mechanisms like that in Fig. 10.5. The turnover in *Alu* copies through such mechanisms is evidently very rapid because, whereas humans have about a million *Alu* copies, chimpanzees have only one-third as many and orang utans have about 600 000. Yet the *Alu* sequences in the haemoglobin gene families in Fig. 10.6 have obviously remained stable for millions of years, since humans diverged from orang utans and chimpanzees. Perhaps *Alu* copies in conserved regions are stable and those in less important stretches of DNA much less stable. *Alu* sequences in themselves perform no function, like other pseudogenes, and seem to change by neutral or mutation-driven evolution; the huge turnover in *Alu* copy numbers between humans and chimpanzees implies that loss or gain is caused by chance, for the functional (coding and control) regions of human and chimpanzee DNA differ by 1% or less and it is hardly possible that those small differences have caused natural selection to favour wholesale accumulation of *Alu* in the human lineage and deletions of *Alu* in the chimpanzee lineage.

Alu sequences are not an isolated example of repeated bits of DNA scattered throughout the chromosomes. The *Alu* family (sequences about 300 bases long) are examples of 'SINEs' (short interspersed repetitive nuclear elements). There is another group of longer repeated sequences called 'LINEs' (long interspersed repetitive nuclear elements), and one family of these (LINE1) has about 100 000 copies in each human nucleus, about 50 000 in chimpanzee and about 80 000 in orang utan. In monkeys, apes and humans the two gamma haemoglobin genes have a LINE1 sequence on either side of them, and between them is a third LINE1, which turns out to be a hybrid of the two flanking LINE1 sequences. This provides an explanation for the gamma

gene duplication in monkeys and apes, by mispairing between the two LINE1 sequences in an ancestral monkey. The ancestor had a single gamma gene (Fig. 10.4), and on either side of it was a LINE1 repeated sequence. In crossing-over there was a mispairing between the two LINEs, resulting in one descendent chromosome with the gamma gene deleted and the other with duplicated gamma genes separated by a hybrid LINE sequence (Fig. 10.7).

10.4 Junk DNA, selfish DNA or ignorant DNA

This and the previous chapter have discussed some of the remarkable recent discoveries about the evolution of DNA. Most these discoveries have been made during the last 15 years. Before that it is fair to say that virtually all biologists were ardent supporters of natural selection – they were 'panselectionists', believing that every feature of every organism is shaped and maintained by function, by selection for the best adapted variants. But DNA does not seem to behave like that, as is obvious from the maxim that *molecular changes that are less likely to be subject to natural selection occur more rapidly*.

The *Alu* sequences discussed above illustrate that maxim, as does the fact that the non-coding parts of genes change more rapidly than the coding parts. Facts like these demand an explanation, and in 1980 two groups of scientists (one including Francis Crick) proposed the theory of 'selfish DNA': their idea was that the 'function' of things like *Alu* sequences is simple self-preservation. We could think of the million copies of *Alu* in each of our cells as rather like the millions of bacteria in our intestine or the millions of copies of a virus that we carry when we have influenza – the environment of the bacteria is our intestine, the environment of the viruses is our cells, and the environment of *Alu* is our DNA. Each (bacterium, virus, *Alu*) will be selected for high rate of reproduction, so building up a large population in its environment. In evolutionary terms, the 'function' of something is that property which causes it to exist. *Alu* sequences have the property of propagating themselves by spreading throughout their environment (DNA), and that property is enough to explain their existence. So the selfish DNA theory sees things like *Alu* sequences as parasites, maintained by natural selection for their ability to propagate themselves without doing too much damage to their host (DNA).

But that idea doesn't seem to work. We have seen that humans and monkeys share a class of identical *Alu* sequences despite the 20 million years separating them whereas humans and apes share *Alu* sequences that are diverging (Fig. 10.6). These facts are best explained if there is a functional and highly conserved gene that is the recent source of the identical human and monkey *Alu* repeats, and the less recent source of the diverging human and ape *Alu* repeats. The only function we can find here is that of the conserved gene – and the dispersed *Alu* copies of part of it are, like other processed pseudogenes, merely by-products of that function that are invisible to natural selection, so that there is no selection between them for reproductive success and they are eliminated (by deletion) or propagated (by duplication) simply by chance. If so, things like *Alu* sequences are not selfish (seeking to propagate themselves), but are just 'ignorant', or 'junk' DNA. However, these repeats must also be 'polite', and not put much of a burden on the host. Humans can survive well enough with a million copies of *Alu* (about 5% of our DNA), but it is doubtful that we could manage with ten million (50% of our DNA).

10.5 Summary

We have seen that the world of DNA is a strange one, far removed from the fit between form and function that seems so obvious in the more familiar world of organisms such as peppered moths, ants and birds. The main things we have learned from the explosive growth of molecular biology in the last two or three decades are:

- that stabilizing or negative selection is the dominant kind of natural selection in the DNA world;
- that neutral mutation and genetic drift are the dominant causes of divergence between the DNA of different lineages; and
- that the DNA of humans (or moths or ants or birds) may sustain large amounts of junk.

But DNA is the only material cause for the successive generations of humans, moths or birds and it is time to get back to the world of organisms, to try to understand the origin of species.

11 The origin of species

We now come back to the problem posed in the title of Darwin's book – can the genetic and environmental factors discussed in the last six chapters account for the division of one species into two, or the appearance of new species? There is one genetic mechanism that produces new species instantly – polyploidy or chromosome-doubling (Section 6.3). This is very common in plants, but not in animals, where we have to find other reasons for species formation.

The fact that has to be explained is the enormous diversity of life. Estimates of the number of species in the world today range from about five million to a hundred million (1.5 million have been named so far). Whichever estimate is closer to the truth, the number of extinct species, known only as fossils, implies that the millions of living species are merely a fraction of those that have existed. If all those species are related by descent, by common ancestry, the development of new species by splitting an ancestral species has happened millions of times – and each split implies change, in one or both descendent lines. This is what we have to explain. Two mechanisms were discussed in earlier chapters:

- natural selection, which will account for adaptive change;
- neutral evolution, which accounts for non-adaptive change in features invisible to natural selection.

11.1 Inconstant environments

The environment is the agent of natural selection, which acts on heritable variations produced by the different types of mutation. Mutation provides the raw material, but selection will propagate a new mutation only if it is favoured by the environment, and this is most likely in a changed or changing environment. This idea of an inconstant environment is one key to species formation. It is comonplace that environmental conditions vary with distance, on a large scale – a winter break in the Canaries will be more enjoyable than one in Spitsbergen – and on a small scale – the daffodils come out earlier in one corner of the garden than in another. We also know that environmental conditions in one place vary with time – Britain was under the Arctic ice sheet in geologically recent times, and most of it was under the ocean not long before that. Information on continental drift (see Fig. 12.8) shows that the concept 'in one place' does not maintain its everyday meaning over long periods – India was once joined to Antarctica and the other southern continents and was under ice, while Antarctica itself was once in the temperate zone.

Knowledge of genetics compels us to think of species not as collections of similar individuals, but as collections of genes or gene pools. These are more or less completely isolated from the gene pools of other species, persist through time and are reshuffled in each generation by the chances of mating to produce new mixtures, which are expressed in the phenotypes of individuals. The fact that each individual requires a certain amount of space makes it unlikely that all individuals of a species will be exposed to exactly the same environment. The larger the population of a species the more space it must occupy, and the more likely it is that environmental changes will be encountered. Within each species new mutations are constantly appearing, and the larger the population the more mutations will appear.

In nature, no species occupies every part of the region through which it occurs. *Homo sapiens*, the most widespread species (except for certain human parasites and microorganisms), is not evenly distributed over the land masses of the world. Large areas – the Arctic, Antarctic, major deserts and mountain ranges – are virtually without human population, and within populated regions people are not

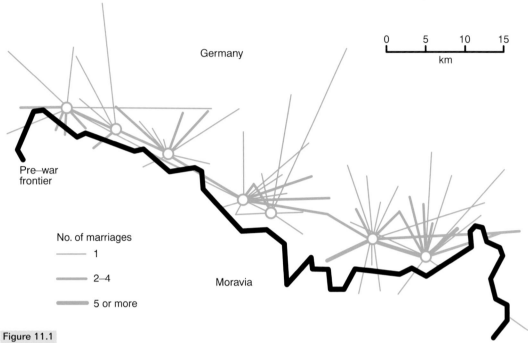

Figure 11.1

Barriers to gene exchange
A political boundary as a barrier to marriage in a human
population. The coloured lines show the direction, distance
and number of marriages made by villagers on the German
side of the pre-war frontier between Germany and Moravia.

equally spread but aggregated into settlements of varying
size, which are separated by thinly populated or uninhabited
areas. Plants and other animals are distributed in the same
way. The original reason for the location of a human settle-
ment may be availability of fresh water, of level ground suitable for planting or of a site offering protection from
attack. The reasons for the patchy distribution of other
species may be harder to guess, but similar factors are
involved. Perhaps a local variation in soil allows a particular
plant species to establish a colony; insects that feed on that
plant will follow, and then species that prey on the insects
will arrive. Even the open ocean, which appears to us as a
uniform mass of water, is broken up into different habitats
by currents, temperature variations, areas of upwelling or
sinking water, and so on.

11.2   Barriers to gene exchange

In the human species each individual may, in theory, choose
as a mate any other individual from the whole population of
the globe. But in practice it does not work out like this.

Even today, when international travel and social mobility
are relatively easy, a person is most likely to marry a close
neighbour, if not the girl or boy next door. Oceans, national
frontiers and even small rivers act as barriers, reducing the
probability of matings across them (Figs 11.1 and 11.2). In
other species the choice of mate is limited by similar factors
– distance and natural barriers between colonies.

Every species is thus broken up into more or less separate inbreeding populations. Chance migrations, temporary
expansions of colonies and other fluctuations make it
unlikely that such populations will be completely isolated
from each other. In genetic terms, a species will be split up
into a number of partially distinct gene pools, with restricted gene exchange (or **gene flow**) between them. Each of
these populations will be exposed to slightly different environmental conditions, and over long periods of time natural
selection may fix different mutations in different populations. Meanwhile, the DNA of these populations will also
inevitably change and diverge purely by chance through the
mechanisms of neutral change (Chapter 9). If interbreeding

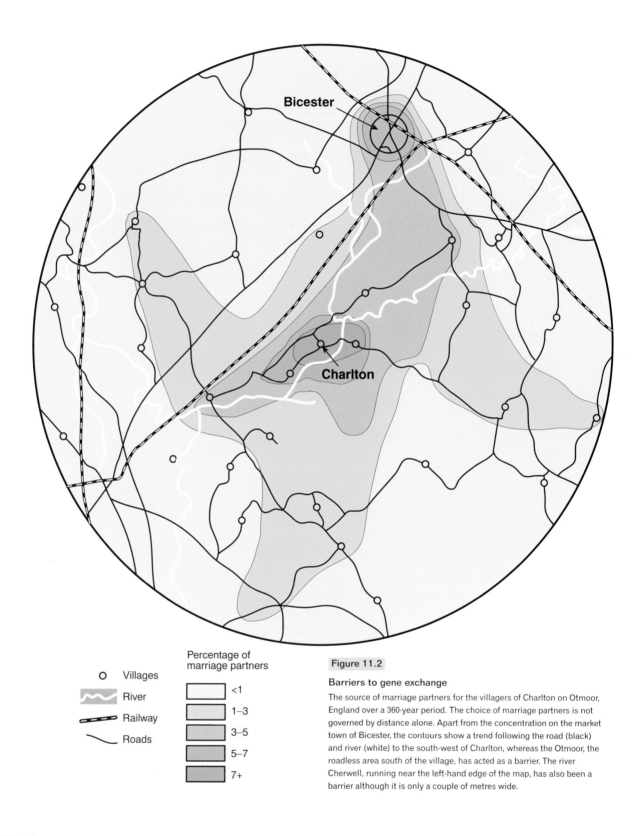

Villages

River

Railway

Roads

Percentage of
marriage partners

<1

1–3

3–5

5–7

7+

Figure 11.2

Barriers to gene exchange

The source of marriage partners for the villagers of Charlton on Otmoor,
England over a 360-year period. The choice of marriage partners is not
governed by distance alone. Apart from the concentration on the market
town of Bicester, the contours show a trend following the road (black)
and river (white) to the south-west of Charlton, whereas the Otmoor, the
roadless area south of the village, has acted as a barrier. The river
Cherwell, running near the left-hand edge of the map, has also been a
barrier although it is only a couple of metres wide.

or gene exchange between neighbouring populations is fairly frequent such local genetic differentiation may not proceed far, but in a widespread species gene exchange between populations at opposite ends of the range will be virtually nil. These terminal populations will never interbreed directly, and will be able to exchange genes only by means of gene flow through interbreeding between the chain of populations linking the ends of the range. There is an analogy in dogs: interbreeding between a great dane and a chihuahua is not possible because of gross disparity in size, but great danes and chihuahuas could exchange genes at one or more removes because they are linked by a series of intermediate breeds which can mate with one another.

In nature, widely separated populations of a species may be exposed to very different environments and may in consequence acquire many genetic differences by natural selection, as well as the inevitable neutral differences that will accumulate over time. The contrast between an Australian aborigine and an Eskimo is an obvious example of this, as are ring-species such as the Herring and Lesser Black-backed gulls described in Section 3.2. The accumulated genetic changes differentiating Australian aborigines and Eskimos do not, so far as we know, impair fertility between the two races in any way. In Britain, the Herring Gull and Lesser Black-backed Gull are the two terminal populations of a circumpolar ring-species. They are interfertile, but breeding between them is so rare that there is virtually no gene exchange. The reason that these two populations remain separate is behavioural. Neighbouring colonies of the two species nest in slightly different habitats, and mate selection by the female gull depends on recognition of males as members of her own species, particularly by the colour of the ring surrounding the eye. Experiments in which eggs of one species are transferred into nests of the other show that this recognition symbol is learnt, by the mechanism known as imprinting. Newly hatched chicks will recognize as their mother the first creature they see, and in adult life will expect their mates to show the same visual signals.

Unlike the mechanisms described in Section 5.1, the barrier separating herring and lesser black-backed gulls is behavioural rather than purely genetic, although the ability to 'imprint' the mother is inborn in chicks and so must also be genetic. However, there is one non-genetic barrier that will prevent hybridization between two populations: **geographic isolation** – the gene pools of Australian aborigines and Eskimos were, until the last couple of hundred

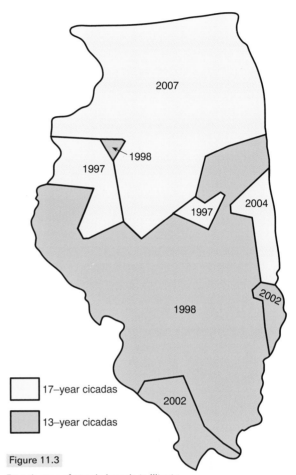

Figure 11.3

Distribution of cicada broods in Illinois
The dates indicate the year of emergence of each brood around the turn of this century.
(After *Biological Notes*, 91, 1975, Illinois Natural History Survey)

years, isolated from one another almost as effectively as those of a fish and an insect. But if two or more species are to coexist in the same place some more positive mechanism must prevent hybridization. Behavioural differences, like those between the two species of gull, can operate only in animals with internal fertilization or fairly elaborate pairing behaviour (such recognition signals need not be visual – for example, difference in song in birds, frogs and grasshoppers may be as effective). In plants, and in animals with simpler courtship, other genetically controlled mechanisms come into play. In addition to those mentioned in Section 5.1 – lack of attraction between gametes, hybrid inviability, hybrid sterility – the two most widespread mechanisms are **ecological** or **habitat isolation**, and **seasonal isolation**.

Habitat isolation is a situation where two populations live in the same region but in slightly different habitats – two plant populations might live at different altitudes or on different soil types; two insect populations might feed and breed in the same place but on different species of plant; two fish populations may live in water of different temperature or salinity. Many examples of this sort are to be found.

Seasonal isolation is the situation where two populations do not interbreed because they mate at different times. The classic example is in periodical cicadas, a group of insects with a very long underground larval stage in the life cycle. In eastern North America there are three species of periodical cicadas, each with two forms, a southern one completing its life cycle in 13 years and a northern one completing it in 17 years. The numbers 13 and 17 were probably not selected at random: both are prime numbers, so that any predator with a shorter life cycle cannot adjust it to coincide with the cicadas at every second or third generation. There are 13 different broods (year classes) of 17-year cicadas, and three different broods of 13-year cicadas. In any one locality a brood of cicadas will emerge only every 13 or 17 years, except in the few places where both forms occur together (Fig. 11.3). Because of the different broods, in the range of 17-year cicadas a brood may emerge in one place in one year and in a nearby place a year or two later. Gene exchange between these broods is prevented as effectively as if they lived thousands of kilometres apart. Where the 13-year and 17-year forms coexist they will meet only every 221 years (13 x 17). Many less spectacular examples of seasonal isolation are known.

11.3 Genetic differences between species, populations and individuals

Failure in hybridization between different species, in nature or in experiments, is an indication that they are reproductively isolated. It can also be looked at as a crude way of estimating the genetic difference between two similar species. Such estimates are more easily made when hybrids can be obtained, especially second-generation hybrids, in which the genes of the two species will combine to throw up many variants. These variations are an indication of the number of genetic differences between the two species. This technique is sometimes possible in plants, in which the gene differences between species that are similar enough to form hybrids are numbered in hundreds rather than in tens.

In animals, where hybridization between species is less common than in plants, other methods of estimating

genetic differences can now be used. The most direct, but at present the most expensive and time-consuming, is to sequence comparable stretches of DNA from individuals of a species and its close relatives. The best example currently available is a stretch of about 17 000 bases of DNA from the beta haemoglobin gene cluster (Section 10.1), sequenced in two humans and in a range of other primates. Most of this is non-coding DNA, presumably invisible to natural selection and changing only by drift or neutral evolution. Over these 17 000 bases the two human sequences show 76 differences, mostly at single bases but with a few small deletions or insertions (Section 6.1), whereas one of the human sequences shows 294 differences from the sequence of the common chimpanzee and the other shows 286. If this fragment can be taken to represent the whole genome, the human sequences (they are from North America) differ from each other by a quarter (average 26%) of the amount that they differ from a chimpanzee. About two-thirds of that long DNA sequence (some 11 000 bases) is also available in a pygmy chimpanzee, representing the second chimpanzee species (the common chimpanzee is *Pan troglodytes*, the pygmy chimp is *Pan paniscus*). Over those 11 000 bases the two human sequences show 58 differences; those two show 244 and 239 differences from the common chimpanzee and 203 and 193 differences from the pygmy chimpanzee; the two chimpanzees show 104 differences. Again, if this fragment is representative of the whole genome, the two chimpanzee species differ from each other by about twice as much (104 vs. 58) as do two humans. These results surprised me when I did the sums after working through the sequences, for the conventional estimate of genetic differentiation within the human species is that it is less than 5% of the difference between humans and chimpanzees (see Fig. 11.5). But that estimate, made before DNA sequences became available, was based on comparisons of proteins and so missed all the variation in DNA at silent sites and in noncoding sequences (Section 7.2). A less extensive comparison is available in the fruit fly *Drosophila*, where a DNA sequence of about 4000 bases can be compared in two individuals from each of three closely related species, and between those three species. The results are broadly similar to the human/chimpanzee comparisons, although the divergences are larger. Within each of the three *Drosophila* species, there are differences between two individuals at about 1.5% of the positions in the sequence (the three figures are 1.2%, 1.6% and 1.7%), and between species the differences are about twice as great (overall average 3.2%;

average differences for the three interspecies comparisons 2.4%, 3.25%, 3.9%). So the divergence within these *Drosophila* species is about half the divergence between species, just as the divergence between two long human sequences is about half that between the two chimpanzee species. I know of no other long DNA sequences from individuals of one species and from its closest relatives, but developments in DNA technology are rapid, and more will soon appear. For the moment, the comparisons are summed up in Fig. 11.4. Two randomly selected human genomes might differ at about 0.5% of the bases in DNA, two fruit fly genomes might differ by about 1.5%, and if we can generalize from humans, chimpanzees and fruit flies, individuals from one species differ at the DNA level by about half as much as do individuals from two closely related species.

A less direct way of measuring variation within and between species is to use DNA hybridization. This estimates the difference between the single-copy DNA (excluding repeated sequences) of two genomes from the lowering of the 'melting point' of DNA in a mixture of the two. In experiments comparing DNA from two humans, the melting point was lowered by 0.3°C, whereas comparisons between the DNA of two humans and four chimpanzees gave an average drop in melting point of 1.6°C. The melting point was lowered by 0.3–0.4°C in comparisons between chimpanzees from the same species, and by 0.8°C in comparisons between common and pygmy chimpanzee. These figures estimate the differences in DNA within species (human or chimpanzee, 0.3–0.4°C) at roughly 20–25% of the difference between humans and chimpanzees (1.6°C), and 40–50% of the difference between the two chimpanzee species (0.8°C). These are close enough to the 25% and 50% estimates that came from the long DNA sequences (Fig. 11.4). There is doubtless experimental error in these estimates, probably more in the hybridization experiments than in the DNA sequences, but error of (say) 10% will not greatly alter the results.

Before DNA sequences became available (during the last decade or so), genetic differentiation within and between closely related species was estimated by comparing many different proteins, to find the proportion with different compositions and therefore different genetic control. If these proteins are regarded as a random sample of the genomes of the organisms, the proportion of protein differences will estimate the difference – the 'genetic distance' – between them. Figure 11.5 summarizes a large number of comparisons made by that method. Although the method catches only differences in DNA that change the coded protein, the comparisons show that we can generalize the results from DNA sequences and DNA hybridization in humans, chimpanzees and fruit flies. Figure 11.5 also shows the genetic distance between human populations and between humans and chimpanzees as estimated from comparisons between proteins, but long DNA sequences (Fig. 11.4 and above) show that the estimate of distance between populations is wildly out. Two human genomes may differ by about 25% of the difference between humans and chimpanzees and by about half of the difference between common and pygmy chimpanzee (sibling species by the categories used in the figure). Two superficially similar species (sibling species) of *Drosophila* differ in thousands, rather than hundreds, of their proteins, but the differences are only about twice as great as those within any one of the species. This reminds us that close physical similarity between two organisms is no guarantee that they have exactly the same genes – one phenotype, or external

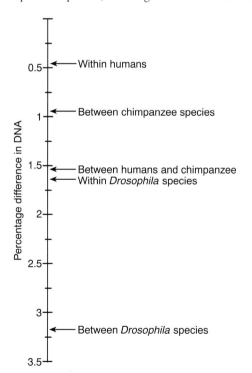

Figure 11.4

Genetic difference between individuals and species

Genetic difference between individuals within a species and in closely related species, estimated from long DNA sequences of humans, chimpanzees and fruit flies.

appearance, can be the product of different genotypes because the phenotype is not the expression of a series of individual genes but a co-operative effort, the result of interaction between genes. It also bears on the origin of species by showing that there is nothing remarkable about the genetic differences between species. Instead, there is a continuum from the difference between two members of the same species, such as humans or *Drosophila melanogaster*, through the difference between members of closely related species, like common and pygmy chimpanzee or sibling species of *Drosophila,* to the difference between more distinct species such as humans and chimpanzees. In a human the half-set of chromosomes received from each parent will differ by about 0.5% of the bases in DNA; as there are about three billion nucleotides in the human genome, that means about 15 million differences between the two sets of chromosomes. Common and pygmy chimpanzees differ by about twice as much, and humans and chimpanzees by about twice as much again. In *Drosophila* the half-set of chromosomes received from each parent differ by about 1.5% but, because *Drosophila* has a relatively small genome, only about two hundred million nucleotides, there will be about three million differences between the two sets of chromosomes and about twice as many between members of two sibling species of fruit fly.

11.4 Summary

Before looking at examples of **speciation** (species forma- tion) it may help to summarize the factors that should influence the origin of new species. In nature, species are broken up into more or less isolated breeding populations. The greater the geographic spread of a species, the greater the range of environments that it will meet. Environmental differences result in high frequency or fixation, by natural selection, of mutations favourable in each set of conditions, and partial or complete isolation of populations means that

they will diverge through neutral changes in DNA, invisible to selection. How far such genetic changes go will depend on gene flow between the different populations, caused by migration or contacts between the populations, and so on. If there is little or no gene exchange over long periods of time, the gene pools of two populations may diverge until fertility between them is impaired when they do come into contact. Reduced fertility can arise from several causes – genetic incompatibility, behavioural, seasonal or ecological isolation. It is important to realize that in sexual species reproductive isolation and barriers to fertile crosses are features of populations, not individuals, and can develop only gradually, by change in the genetic constitution of the whole population. Isolating mechanisms, unlike mutations, cannot arise in one individual because it would be prevented from breeding (except by self-fertilization) as effectively as if it had developed a lethal mutation.

We should therefore expect new species to result from long-continued isolation of different populations of a species. The most obvious examples are when populations of a terrestrial species live on different islands. Rather than select examples from a variety of places we will look at one group of islands – the Galapagos – in some detail.

Figure 11.5

Estimates of genetic distance

Genetic distance in groups of animals, estimated from comparisons of many proteins in insects, crustaceans, fish, amphibians, reptiles and mammals. The bars show the range in groups with different degrees of relationship. 'Populations' are different geographic groups within one species. 'Semispecies' are populations that may hybridize, like members of the same species, but which show behavioural and other differences that limit interbreeding (like the gulls discussed on page 81). 'Sibling species' behave like different species and do not interbreed, but are so similar that they cannot be distinguished easily (like the *Anopheles* mosquitoes discussed on page 3). 'Congeneric species' are members of the same genus, such as lion and tiger, or wolf and coyote. The numbers on the scale are roughly equal to the number of non-silent mutations fixed in each gene.

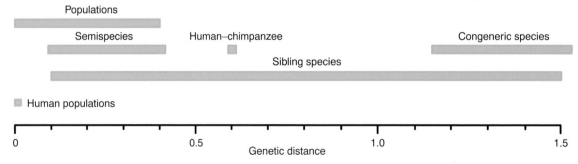

12 Speciation in the Galapagos Islands

In July opened first note book on 'Transmutation of Species' – Had been greatly struck from about Month of previous March on character of S. American fossils – & species on Galapagos Archipelago. These facts origin (especially latter) of all my views.
Darwin's Journal, 1837

In 1835, Darwin spent five weeks in the Galapagos archipelago (Fig. 12.1), in the eastern Pacific, during the voyage of HMS *Beagle*. His experiences in the islands were to be crucial in convincing him that species can change and divide. The *Beagle* had been surveying the coasts of South America for more than three years, and during frequent trips ashore Darwin became thoroughly familiar with the fauna and flora of the continent, so that he was prepared for the peculiarities of Galapagos animals and plants.

The Galapagos archipelago is an isolated group of volcanic islands lying on the equator, about 1000 km east of the South American mainland. The largest island, Isabela, is about 120 km long, and its highest volcano reaches 1650 m. There are four other fairly large islands, 11 smaller ones, and many small islets and rocks. The large islands have quite a rich flora, zoned according to altitude, with dense humid forest between about 220 and 440 m. The original fauna of the islands has been ravaged by introduced animals such as goats (first released in 1813), donkeys, pigs, cats, rats, ants and cockroaches, but in Darwin's time not too much damage had been done. Darwin and other members of the *Beagle's* crew went ashore on only four islands: San Cristóbal (Chatham), Santa Maria (Charles), Isabela (Albemarle) and San Salvador (James). The islands were made a national park by the government of Ecuador in 1936, and in 1959 (the centenary of publication of *The Origin of Species*) the international Charles Darwin Foundation for the Galapagos Islands was set up. The Foundation now maintains a research station on San Salvador.

12.1 Fauna of the Galapagos

The fauna of the Galapagos, though very peculiar, is clearly related to that of the South American mainland. Darwin wrote 'it was most striking to be surrounded by new birds, new reptiles, new shells, new insects, new plants, and yet by innumerable trifling details of structure, and even by the tones of voice and plumage of the birds, to have the temperate plains of Patagonia, or the hot dry deserts of Northern Chile, vividly brought before my eyes'. He went on to say that the islands are quite different from the South American coast in geological and climatic features, but are very like the Cape Verde islands in the Atlantic. Yet the Galapagos fauna and flora resemble those of South America, not the Cape Verde islands, which have animals and plants of African type.

Reflecting on his time in the Galapagos, Darwin emphasized one fact: 'by far the most remarkable feature in … this archipelago … is that the different islands … are inhabited by a different set of beings … I never dreamed that islands, about fifty or sixty miles apart, and most of them in sight of each other, … would have been differently tenanted.' This was the judgement of hindsight, written after he got back to London and worked up the collections made in the archipelago. Darwin might first have appreciated that the islands are 'differently tenanted' when the Vice-Governor of the Galapagos told him that he could tell from which island any giant tortoise had come by features of its shell. But at that time Darwin believed that the tortoises belonged to a species that existed on oceanic islands in many parts of the world, and so were probably introduced, not native. He collected many specimens of plants and animals, notably of a group of land birds now called 'Darwin's finches' and known to vary from island to island, but he kept no record of which bird came from which island. When Darwin reached London he sent his bird collections to John Gould, the ornithologist at the London Zoo, and was astonished to learn from Gould that nearly every Galapagos bird represented a new, endemic species.

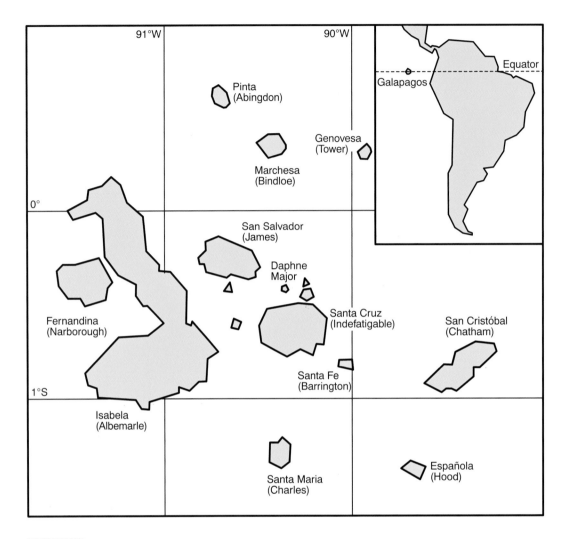

Figure 12.1

Map of the Galapagos archipelago

The archipelago was annexed by Ecuador in 1832, and official Spanish names were given to the islands in 1892. The old English names of the islands, used in Darwin's time and in the names of many Galapagos species, are given underneath the current Spanish names.

In particular the Galapagos finches comprised a unique cluster of 13 new species. Darwin (who had mistaken one of those finches for a wren and another for an oriole) had to find an explanation for these surprising facts, and so began to think about 'transmutation of species'. Since Darwin's time the islands have been carefully studied by naturalists, and his opinions on the significance of the animals and plants have been fully confirmed. Here there is space to discuss only a few of the animals – and the reptiles and birds provide good examples.

Looking first at the fauna of the archipelago as a whole, without distinguishing particular islands, the assemblage of animals is unusual. Until domestic animals were introduced there were no mammals except for rats, bats and seals.

Land and water birds are present in some variety and include the northernmost colonies of penguins, but many important South American types are missing.

Galapagos reptiles include giant tortoises, a few snakes and many lizards, including large herbivorous iguanas. There are no amphibians (frogs, newts, salamanders, etc.) and no freshwater fishes. Most Galapagos insects are small, drab and retiring. There is only one type of bee, few butterflies or moths, and no native aphids, fleas or representatives of many other insect groups.

12.2 Galapagos reptiles

The reptiles are amongst the most abundant and interesting Galapagos animals. Taking the giant tortoises first, the opinion of the Vice-Governor in Darwin's time – that each island has its own variety – has been confirmed. The tortoises are the largest native land animals, up to 2 m long and 270 kg in weight. Galapagos is an old Spanish word for tortoise, and the tortoises attracted early visitors – buccaneers and whalers – to the islands in search of fresh meat. Today, tortoises are seen on seven of the islands. On three others they have been exterminated, by humans, within the last 150 years. All these tortoises belong to the same species, *Geochelone elephantopus*, but 14 subspecies are recognized, differing in shape, colour and thickness of shell, length of neck and legs, and in size. Nine islands each have their own peculiar subspecies (three of these are now extinct, and on Pinta there is only one survivor, known as 'Lonesome George'). The largest island, Isabela, has five different subspecies, each confined to the neighbourhood of one of the five major volcanoes on the island. Comparison of proteins (by electrophoresis) from seven of the subspecies showed that the most distinctive (and so, presumably, the oldest) is that from San Cristóbal, followed by those from Española and Pinta; the four subspecies from the central group of islands were all closely clustered. A similar comparison with a range of mainland species showed that *Geochelone elephantopus* is not closely related to any of them, and so has been distinct for millions of years.

There are two main types of Galapagos tortoise, one with a domed shell and a short neck (Fig. 12.2(a)), the other with long legs and neck and a shell that flares up in front, so that the head and neck can be raised (Fig. 12.2(b)). The distribution of these two types is associated with habitat – the long-necked forms live on arid islands with much broken ground, where they must raise their heads to feed on shrubs and prickly pear; the short-necked

(a)

(b)

Figure 12.2

Galapagos giant tortoises, *Geochelone elephantopus*

(a) A short-necked, domed-shell form from the north of Isabela Island: an adult male, about 1 m long. (b) A long-necked, flared-shell form from Pinta Island: two views of an adult male (about 1 m long).

(a)

(b)

(c)

(d)

Figure 12.3

Galapagos iguanas
(a) Land iguana,
Conolophus subcristatus,
drawn from a specimen
brought home by Darwin,
in *Zoology of the Voyage of
the Beagle* (1841).
(b) Land iguana,
Santa Cruz Island.
(c) Marine iguanas, Male
Amblyrhynchus cristatus,
displaying on the shore of
Fernandina Island.
(d) Marine iguana,
Española Island.
(Photos by
Dr Andrew Milner.)

varieties live on moister islands, where grass and other low foliage is available as food. So it seems that some of the differences between the island races are adaptive.

The prickly pear cactus, the principal food of the tortoises, also varies from island to island. On islands where tortoises have never occurred the cacti are low and spreading, with soft spines, but on all the islands inhabited by tortoises the cacti are erect and tree-like, with stiff spines. The tree-like cacti (four different species) seem to have evolved this growth habit in response to browsing by tortoises.

The large iguanas are perhaps the oddest and most characteristic of the Galapagos reptiles. There are two sorts: land iguanas (*Conolophus*) and marine iguanas (*Amblyrhynchus*) (Fig. 12.3). Both are large animals, reaching over 1 m in length. They are related to iguanas on mainland South America, but are each other's closest relatives; estimates of the genetic difference between the land and marine iguanas suggest that the two lineages have been distinct for 15–20 million years.

Land iguanas live in burrows and are mainly herbivorous, although they also eat insects. They are found only on the central group of islands, and there are two very similar species, one restricted to Santa Fe island, the other originally present on five islands, but now extinct on two of them. These two island populations have been exterminated in the last century, one by competition from introduced pigs, the other following the establishment of an American air-base during the Second World War.

Marine iguanas are unique to the Galapagos. These animals spend most of their time on the shore, and enter the sea only to feed. They eat seaweed, clinging to rocks in the surf or swimming down to the bottom in depths of as much as 11 m, where they can stay for nearly an hour without surfacing to breathe. They differ from the land iguanas in having partially webbed toes and a deep, flat-sided tail, used in swimming. They have been seen up to a kilometre out at sea, but rarely go so far from land. Marine iguanas occur on all the islands, and all these populations are included in a single species. But there are differences in island populations, especially those on the remote southern and northern islands, in colour, behaviour and size.

Small lizards, called lava lizards (*Tropidurus*), are found on all the islands except three remote northern ones (Fig. 12.4(a)). There are seven species of these lizards, differing in colour, size, shape, scale pattern and so on, and on each of the 12 islands where lizards are found there is only one species. *Tropidurus albemarlensis* occupies the central group

of four large islands and two nearby small ones. Each of the other six species is confined to one island. Amongst the six different island populations of *T. albemarlensis* the display behaviour is characteristic for each island, indicating some evolutionary divergence, and this is confirmed by estimates of genetic distance between the populations.

Tropidurus lizards also occur on the South American mainland. Comparison between proteins from Galapagos and mainland species indicates that two Galapagos species, *T. bivittatus* from San Cristóbal and *T. habelii* from Marchesa, are related to each other but are very different from the other five Galapagos species. *Tropidurus delanonis*, from Española, is also quite distinct. Further, *T. bivittatus* and *T. habelii* seem to be closer to a mainland species from coastal Ecuador than they are to other Galapagos species (Fig. 12.4(b)). As with the land and marine iguanas, it seems that some of the Galapagos lava lizard lineages have been distinct for up to 15 or 20 million years, and if a mainland species is the closest relative of *T. bivittatus* and *T. habelii* it would follow that there are at least two historical links between Galapagos and mainland lizards.

12.3 Galapagos birds

Galapagos birds include several species that are found elsewhere and which show no local differentiation in the archipelago. This category contains sea birds (terns, petrels, frigate-birds, etc.) and some land birds (cuckoo, moorhen). Other widespread species have developed a local subspecies in the Galapagos – a pelican, a tern, two herons, a duck, a flamingo and two owls. In other cases the Galapagos form is sufficiently distinct to be named as a local species, as is the albatross, the penguin, a gull and the hawk. Some Galapagos birds are so different that they are placed in a special genus (Section 13.1), such as the peculiar flightless cormorant, a nocturnal gull and the Galapagos dove. None of the birds mentioned so far has evolved island varieties within the archipelago.

Local differentiation is found especially in two groups of birds:

- the mocking birds, a Galapagos genus with four species;
- Darwin's finches, distinct at the tribe or subfamily level and containing five genera and a total of 14 species.

The four species of mocking birds (thrush-sized, predatory birds) are distributed in the same way as some of the reptiles – no island has more than one species, three of the species are each confined to one outlying southern island (that on Santa Maria is now extinct but survives on two

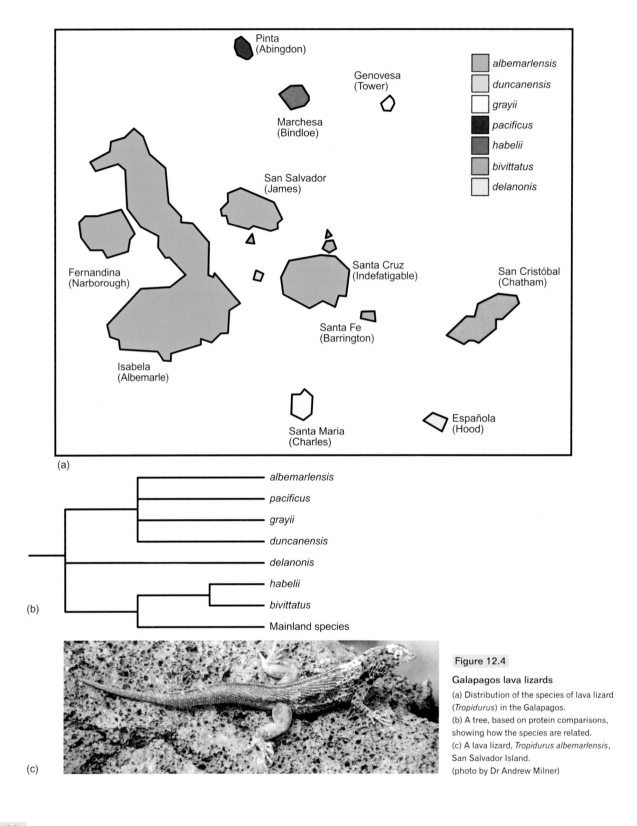

(a)

(b)

(c)

Figure 12.4

Galapagos lava lizards

(a) Distribution of the species of lava lizard (*Tropidurus*) in the Galapagos.
(b) A tree, based on protein comparisons, showing how the species are related.
(c) A lava lizard, *Tropidurus albemarlensis*, San Salvador Island.
(photo by Dr Andrew Milner)

(a)

(b)

Figure 12.5

Darwin's finches
(a) A Tree Finch, *Camarhynchus psittacula*. It has a short, thick beak and feeds mainly on beetles and similar insects taken from the bark of trees.
(b) The Cactus Ground Finch, *Geospiza scandens*, has a long, slender beak and feeds mainly on the flowers of prickly pear cactus. Natural size, from John Gould's plates in *Zoology of the Voyage of the Beagle* (1841).

nearby satellite islets), and the fourth occurs on all the other islands and has seven subspecies, most of the remote islands having one subspecies each. Their closest relative is a single species in western South America; although it is placed in a separate genus, a hybrid (in captivity) between that species and one of the Galapagos species has been reported.

Darwin's finches are the most common land birds in the Galapagos. They are mostly sparrow-sized, and are all dull in colour, the females grey–brown and the males mostly black. The different genera and species are recognized mainly by differences in size, in the size and shape of the beak and in feeding habits (Figs 12.5-12.7).

- One group, the ground finches (*Geospiza*), contains six species which live in the coastal zone and the lowlands, and usually feed on the ground, on seeds and insects. The species differ in the shape of the beak and in the size and toughness of the seeds on which each specializes. One species has a long, curved beak and feeds on the flowers of prickly pear cacti.

- A second group, the tree finches (*Camarhynchus*), contains three species with deep, parrot-like beaks. They live in the forest zone, feeding in the trees on insects and seeds.
- The Warbler Finch, placed in a genus of its own, has a slender beak and feeds only on insects, sometimes taken on the wing.

Large Ground Finch
Geospiza magnirostris

Medium Ground Finch
Geospiza fortis

Insectivorous Tree Finch
Camarhynchus parvulus

Warbler Finch
Certhidea olivacea

Figure 12.6

Darwin's finches

The heads of four species of Darwin's finches.
From Charles Darwin's *Journal of Researches* (1845).

- Another species, also usually placed in a genus of its own, is the Vegetarian Tree Finch, a larger bird living in the forest zone and feeding on fruit, buds and soft seeds.
- Finally, two other species, sometimes placed in a fifth genus, have habits like woodpeckers, feeding on insects taken from tree trunks and branches. Unlike woodpeckers they do not have strong, sharply pointed beaks and long tongues, but they make up for this by using twigs or cactus spines as tools to poke insects out of crevices. One of these birds, the Mangrove Finch, ranges up into the forest zone.

The nearest mainland relative of Darwin's finches is not known, but a likely candidate is the Blue–black Grassquit (*Volatinia jacarina*), now found in Central and South America and in the West Indies. The only other member of the Darwin's finch group is a species on Cocos Island, 720 km north-east of the Galapagos; this Cocos Island finch is closest to the Galapagos Warbler Finch, which is the most distant member of the Galapagos cluster (Fig. 12.7). Comparison of proteins from 11 of the 13 species of Galapagos finches indicates that the genetic differences between the genera and species (except for the Warbler Finch) are small compared with other birds, suggesting that most of the species have diverged rather recently, perhaps within the last million years or so.

Darwin's finches show a wide range of feeding habits. Their distribution in the archipelago is unlike that of the reptiles and mocking birds, where each island has only one species or subspecies of a particular animal. Each island usually has several species of finch, and there are 11 islands with seven or more species. Only one species is confined to a single island; one other species is found only on two islands. The finch fauna of an island will contain representatives of different groups of finches rather than several species of the same type.

The ground finches and the Warbler Finch are the most widespread, occurring on all the islands. Tree finches and the Vegetarian Tree Finch are missing from the outlying islands to the south and north, and the tool-using finches are restricted to the central group of islands. Within the more widely distributed species there is local differentiation of island subspecies or races, differing in song, behaviour and other details.

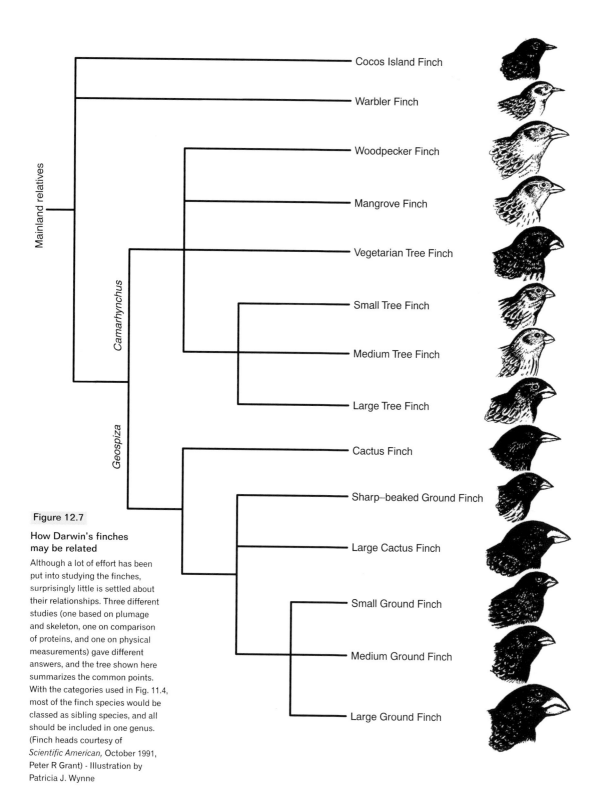

Mainland relatives

Cocos Island Finch

Warbler Finch

Woodpecker Finch

Mangrove Finch

Vegetarian Tree Finch

Camarhynchus

Small Tree Finch

Medium Tree Finch

Large Tree Finch

Geospiza

Cactus Finch

Sharp–beaked Ground Finch

Large Cactus Finch

Small Ground Finch

Medium Ground Finch

Large Ground Finch

Figure 12.7

How Darwin's finches may be related

Although a lot of effort has been put into studying the finches, surprisingly little is settled about their relationships. Three different studies (one based on plumage and skeleton, one on comparison of proteins, and one on physical measurements) gave different answers, and the tree shown here summarizes the common points. With the categories used in Fig. 11.4, most of the finch species would be classed as sibling species, and all should be included in one genus. (Finch heads courtesy of *Scientific American,* October 1991, Peter R Grant) - Illustration by Patricia J. Wynne

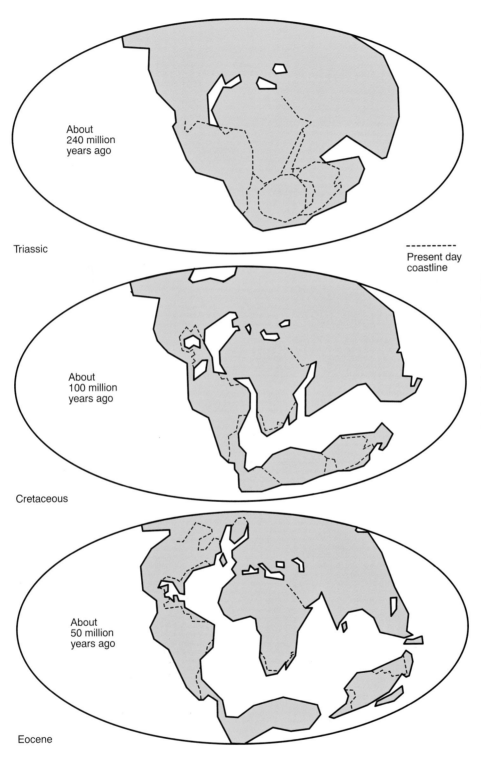

Triassic

About
240 million
years ago

Cretaceous

About
100 million
years ago

Eocene

About
50 million
years ago

- - - - - - - - - -
Present day
coastline

Figure 12.8

Continental drift

Sketch maps showing
the distribution of the
continents at three
different times. The
decisive evidence for
drift, and for the
former positions of the
continents, comes from
the discovery that the
floor of the oceans
consists of strips which
are magnetized in
alternate directions,
reflecting periodic
reversals in the earth's
magnetic field. These
strips are symmetrically
arranged on either side of
mid-oceanic ridges, along
which new crust is welling
up. By correlating the
magnetic reversals with
events of known age, the
position of the continents
at different times can be
estimated by subtracting
younger ocean floor. These
maps are drawn according
to the theory that the Earth
is expanding: the Triassic
globe is about 15% smaller
than the Eocene one. This
theory is controversial but
it gives a better fit for the
continents than does a
globe of fixed diameter.

A careful, long-term study of the finches has been made on Daphne Major, an islet north of San Salvador with an area of only about 0.3 km². On Daphne Major the Medium Ground Finch (*Geospiza fortis*) and the Cactus Finch (*G. scandens*) are common and there are a few Small Ground Finches (*G. fuliginosa*); average breeding populations are about 200 *G. fortis*, 100 *G. scandens* and six *G. fuliginosa*. The study produced two surprising results.

- Both *G. scandens* and *G. fuliginosa* hybridize naturally and successfully with *G. fortis*. Over 12 years, about 2% of breeding *G. fortis* were in hybrid pairs, and individuals of *G. fuliginosa* bred more often with *G. fortis* than with their own species, presumably because there were so few of them.
- Both kinds of hybrid pairs were more successful than pairs of the same species, in both survival and fertility of offspring. In other words, the hybrid offspring are fitter (page 46).

Although *G. scandens* and *G. fuliginosa* do not hybridize there is occasional breeding between *fortis/scandens* and *fortis/fuliginosa* hybrids, so the two species can exchange genes. The result of these patterns, over many years, should be that the populations of the three species will eventually fuse into one interbreeding population. But the scientists who carried out the study believe that outcome to be unlikely because hybrids will be at a disadvantage in years of drought or flood. Long-term studies of *G. fortis* have shown that large individuals (with large beaks) are favoured by natural selection in drought whereas small individuals are favoured in wet conditions, because different seeds are available as food. Hybrids are intermediate between their parent species in beak size, and so will be selected against when times are hard.

12.4 Galapagos history and species

Before summarizing the evidence from Galapagos reptiles and birds on the process of speciation, it is worth considering how and when the archipelago became populated. As volcanoes, some of them still active, the islands evidently came out of the sea comparatively recently and would have had no flora or fauna when they first emerged. When the continents were thought to be fixed (until the 1960s) it was natural to assume that the Galapagos and the American mainland had been in the same places since the islands first appeared, and that animals and plants reached the islands from the mainland by rare accidents. The South Equatorial Current flows from the South American coast to the islands and could have carried floating objects – seeds, animals, rafts of vegetation. Galapagos tortoises float, and some have survived several days at sea. Lizards and snakes or their eggs might be transported on tree trunks or mats of vegetation. The trade winds also blow from the mainland to the islands, and might help stray birds on their way; during the severe El Niño event of 1982–1983, when the islands experienced torrential rain and high winds for months, individuals of at least three species of small mainland birds were recorded for the first time in the Galapagos. The traditional view is that the islands were colonized like this, by the chance arrival of perhaps a single pair of one species, or a single fertilized female of another.

But, according to the theory of **continental drift** (Fig. 12.8), or **plate tectonics**, the crust of the Earth consists of a number of rigid plates, some of them bearing continents, which are in constant motion relative to one another, and may grow by the generation of new crust at mid-ocean ridges or shrink by the consumption of old crust at ocean trenches. Two geographic features will be in relative motion unless they are on the same plate.

The Galapagos islands lie close to the junction of two plates, the Nazca and Cocos Plates, in an area of crust generation and complex motion (Fig. 12.9). The islands are the product of a volcanic hot-spot at this junction. The hot-spot has been active for at least 30–40 million years, but the existing islands are less than five million years old. The oldest rocks are found on the eastern islands, San Cristóbal and Española, and are about three million years old. The western islands, Isabela and Fernandina, are much younger, half a million years or less. The existing islands are evidently only the latest of a series of archipelagos, successively erupting, moving eastwards from the hot-spot, and being eroded away. Traces of earlier islands, from which rocks 5–9 million years old have been dredged, exist in a chain on the sea bottom for up to 600 km to the east of the islands, on the Carnegie Ridge, which extends as far as the mainland and marks the past track of the Nazca plate over the hot-spot. Even older islands, now completely eroded away, presumably existed on this ridge and may have been closer to the mainland. To the north-east of the Galapagos a second submarine ridge, the Cocos Ridge, extends to central America and marks the track of the Cocos plate over the hot-spot. On the Cocos Ridge lies Cocos Island, the home of the only close relative of Darwin's finches.

So, although we cannot say exactly how or when the present inhabitants of the islands arrived, it is clear that

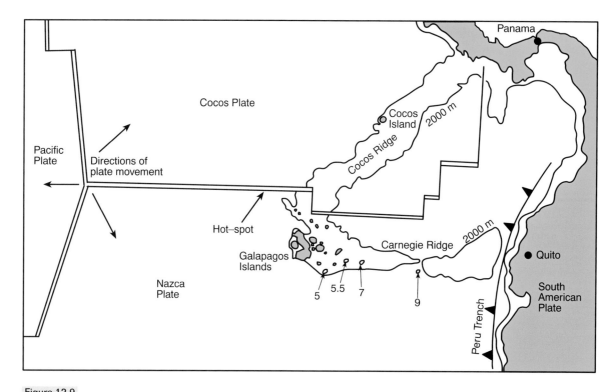

Figure 12.9

History of the Galapagos hot-spot

The Peru Trench is an area where oceanic crust is being consumed, sinking beneath the
South American Plate. The Cocos and Carnegie ridges mark the track of the Cocos and
Nazca plates over the hot-spot, and dots along the Carnegie Ridge show sites where
rocks have been dredged from drowned islands, with their ages in millions of years.

present geography is not the best guide. The existing
Galapagos islands are young (less than five million years),
but they are only the most recent occupants of a spot that
has been generating rocks for at least 15–20 million years,
and probably for much longer. Darwin's finches may be the
descendants of a single pair of birds blown, by chance, from
the mainland; or they may be the descendants of one or
more finch populations on some earlier island or islands,
possibly closer to the mainland.

Luckily it is not necessary to know how or when each
element of the present Galapagos fauna arrived before we
can draw some inferences, of general validity, about how
new species arise. The water between the archipelago and
the South American mainland is an effective barrier, as is
shown by the absence of amphibians and native large mammals – although there are habitats suitable for both (introduced pigs and donkeys have thrived). The water between
the islands is also an effective barrier – even to some birds,
which are physically quite capable of the crossing but do
not occupy all islands with suitable habitats. In the archipelago as a whole some species (insects, birds, reptiles) are the
same as on the mainland; others are differentiated into
Galapagos subspecies or species, or are sufficiently divergent to be placed in a Galapagos genus. The same effects are
seen within the archipelago – related populations on different islands may be different species (as in the lava lizards
and mocking birds), but the giant tortoises have diverged
only slightly, as subspecies, and the land iguanas have hardly
diverged at all on most islands. So isolation does not invariably produce new species, but we cannot say if this is
because some island populations have not been separated
long enough, or because chance migration from island to
island maintains gene exchange between populations, or
because the islands are so similar that no adaptive change
is selected, or if yet other factors are involved. In general,
age and remoteness of an island seem to correlate with

differentiation of its inhabitants, as with the tortoises, which are most distinct on San Cristóbal and Española (the two oldest islands), and the lava lizards, most distinct on those two islands and on the northern Marchesa. The greater uniformity of the fauna of the central group of islands could be an indication that they were linked by dry land in quite recent times, until separated by subsidence or changes in sea-level.

Some of the divergence between island forms is obviously adaptive, as in the long-legged and long-necked tortoises or the differences between land and marine iguanas, and especially in the adaptations of the different groups of finches. These finches are a classic example of **adaptive radiation**, where one type of animal or plant includes a variety of forms adapted to different ways of life. Why have the finches of South America, or of Great Britain for that matter, not developed a range of adaptations like the Galapagos finches? The reason must be that British finches are unable to produce warbler-like or woodpecker-like forms because those ways of life, or **ecological niches**, are already filled – by warblers and woodpeckers. This emphasizes the fact that an important agent in natural selection is competition between different species for the limited resources of the environment. The same point is illustrated by the adaptations of Galapagos tortoises and prickly pears – on islands where tortoises occur, filling the niche of large herbivores (occupied on the American mainland by large mammals), the prickly pears have responded by developing stiff spines and a tree-like form.

The finches raise another aspect of speciation. They are almost the only Galapagos animal group in which one island may contain several species. Is it possible that these different species diverged on one island? This question has led to long discussions – whether geographic isolation of populations is always necessary before species can diverge, or whether they can split in one place. It is not necessary to go into the details of these arguments here. The problem is analogous to the spread of an advantageous mutation, discussed in Section 7.3. Suppose, to take an extreme example, that one member of a finch population hit on the idea of using a thorn to dig insects out of tree bark. Its offspring could pick up the same trick, and in turn might teach their offspring. Could a new species, in which the tool-using habit was genetically fixed, arise in this way? Only if the first tool-using finches were much more likely to mate with birds having the same habit. This might come about if the tool-using finches fed and mated in the trees, while the

ancestral non-tool-using population fed and mated on the ground, perhaps in more open country. But this is, in effect, geographic isolation, though on a small scale.

Nearly all supposed examples of speciation within an originally undivided population seem to hinge on small-scale geographic (or ecological) isolation like this. So speciation in Galapagos finches could, in theory, have been initiated within one island population. But the examples of mocking birds and the various reptiles, where one island always has only one species or subspecies, make it more probable that the finch species diverged in the same way, on different islands. If so, the present situation (several species of finch on one island) is due to reinvasion, each species spreading through the archipelago from its original island home. The fact that two or more species of finch can co-exist in the same habitat indicates that they have developed effective reproductive isolation, preventing gene exchange. Observation of ground finches on Daphne Major (Fig. 12.1) shows that hybridization between some species occurs and is successful, but there it is supposed that hybrids are selected out in years of drought or flood so that distinct species are maintained by natural selection. Co-existence between finch species on the same island is also an indication that they have different habits, because if they had exactly the same ecological requirements they would be in direct competition, and one would surely be more efficient at finding those requirements than another, which would eventually become extinct.

There is one more general question about the origin of species. In choosing the Galapagos as an example, and concentrating on the land animals there, I have cheated. For the environment open to these animals is broken up into separate islands, with new islands appearing and old islands being eroded away over a few million years. In that fragmented and changing geography it is easy to imagine how two new species may develop, through isolation and divergence, on different islands. Looking back to the origin of these two island species, we could guess at two possibilities.

- In the first the animals lived only on one island until by chance two or a few individuals reached the second island: the new species represents the descendants of these colonizers.
- In the second the animals lived on one island, which, through changes in geography or in sea level, became subdivided into two: the new species represent the descendants of a population divided by natural processes. Which is the more general model of how new species

Figure 12.10

Alternative models of the origin of new species

The shapes represent the entire population of a species, with time indicated by upward growth. (a) A species splits into two roughly equal daughter-species, which gradually diverge and differentiate. (b) A small population differentiates rapidly and develops into a successful new species while the parent species remains unchanged. In this second model we should imagine that the 'trunk' is continually budding off 'twigs' – small, divergent populations – most of which die out or merge back into the 'trunk'.

originate? On a volcanic archipelago we might prefer the first: it is the animals that move. But in a more uniform environment, like the ocean or the American prairies, we might choose the second, in which a large population was passively divided by the slow development of a new geographic or ecological barrier.

These two alternative models are shown diagrammatically in Fig. 12.10. In 12.10(a) the process of speciation is gradual, involves large populations, and there is no decisive moment or event in the origin of the two new species. In 12.10(b) the parent species remains unchanged, and the new species originates rapidly following a particular event, like the arrival of a founder population of lizards on a new island or of insects on a different species of food plant.

Under these conditions – a new environment and a small population – change might be very rapid. The new environment will exert new selective forces or, if it is uninhabited, selection may be relaxed as the population expands to fill the empty niche. The small original population will mean inbreeding, and so rapid fixation of genes, by natural selection or by genetic drift. In these circumstances a 'genetic revolution' could occur and a new species, infertile with the parent species, might develop in a few generations.

These two models of speciation are not mutually exclusive. The gradual model (the one favoured by Darwin) might prevail in rather uniform environments, and the rapid, revolutionary one (sometimes called **quantum speciation**) in fluctuating or 'patchy' environments.

Intermediates between the two models can be imagined. Examples can be found to fit each model, and each has its devotees. But the fact is that no one has actually observed the origin of even one new species in nature and we cannot tell if the gradual or quantum model is the dominant one. Nor can we evaluate the title of Darwin's book (*On The Origin of Species by Means of Natural Selection*) for we do not yet know whether the dominant force in speciation is Darwin's natural selection, or the neutral changes that will inevitably accumulate to differentiate populations that are separated.

A final point about Darwin's finches, the most diverse Galapagos group, is that they illustrate what might be called 'rounds of speciation'. Their diversity exists on at least three levels:

- between the main groups (now classified as genera) such as ground finches and tree finches;
- within each group, as between the six species of ground finch (genus *Geospiza*);
- within each species, as between the island races of a single widespread ground finch.

These three levels would be the product of three successive, and continuous, rounds or episodes of divergence, initiated by the splitting of one original species. This idea, of a hierarchy of species splits, is explored further in Chapter 13.

13 The hierarchy of life

I cannot doubt that the theory of descent with modification embraces all the members of the same great class or kingdom.

Charles Darwin

The Galapagos archipelago gives many examples of speciation – pairs or clusters of closely related species that have diverged relatively recently from an ancestral species. Equally good examples can be found in other parts of the world: in Britain and its offshore islands, in the sea on either side of the isthmus of Panama, and so on. But all such examples concern change and divergence on a relatively trivial level. It is reasonable to ask whether small-scale changes like these can possibly account for large-scale differences like those between a human and a mouse, or an elephant and an oak tree. And could the accumulation of small mutations in DNA account for the perfection and complexity of a peacock's tail, or a hawk's eye or the human brain?

There are three main lines of approach to these questions:

- the classification of animals and plants
- fossils and the geological history of the world
- the newer field of molecular evolution.

13.1 Classification

And thus, the forms of life throughout the universe become divided into groups subordinate to groups.

Charles Darwin

Darwin found one of the strongest arguments for his theory in the fact that animal and plant species fall into groups and that these groups form a nested series or hierarchy, smaller groups or sets of species (such as owls and ducks, or seals and deer) being included within successively larger groups (birds and mammals within vertebrates; vertebrates, insects and molluscs within animals, etc.). Observations like this have certainly been made since antiquity, but the system of classification now in use in biology is based on the work of the eighteenth-century Swedish naturalist Linnaeus.

Linnaeus introduced binomial names, giving each species two Latin or latinized names, of which the first (genus name) is a group name which may be shared by several species and the second (trivial name) is particular and identifies the individual species. For example, the Linnaean genus *Turdus* contains several British and European species, including *Turdus merula* (the blackbird), *T. philomelos* (the song thrush), *T. viscivorus* (the mistle thrush), *T. pilaris* (the fieldfare) and *T. torquatus* (the ring ouzel). *Turdus* is represented in other parts of the world by many other species (about 60 in all), including the American robin, *T. migratorius*. Just as species are grouped in a genus, genera may be grouped in a set of higher rank, the family (*Turdus* is grouped with true robins, nightingales, redstarts, etc. in the subfamily Turdinae, which in turn is placed in the family Muscicapidae, including another subfamily for warblers, flycatchers and tits), families may be grouped in orders (the order Passeriformes contains all perching birds), orders in classes (the class Aves contains all birds), classes in a phylum (e.g. phylum Vertebrata), and phyla in a kingdom (e.g. kingdom Animalia).

When he introduced this system, Linnaeus intended it partly as a convenient aid to the memory, a means of making comprehensible the diversity of nature. He recognized about 12 000 species of plants and animals, grouped into about 1000 genera, and expected the educated man to be able to remember the genera (today species are numbered in millions, and genera in hundreds of thousands). From this point of view – convenience – the Linnaean system hardly differs from other hierarchical classifications, such as those used in libraries where books are divided by topics and subtopics, each subgroup is further broken down by size and then by alphabetical order of authors. However, Linnaeus also had a higher purpose than merely to catalogue nature. He believed that he was uncovering the plan of the Creator, since he thought that species were fixed and individually created. (Later in life Linnaeus seems to have decided that species might change and that God could originally have created only the genera.)

Linnaeus and his successors recognized genera, families and other categories on the basis of similarities in structure, and believed that each group had a set of features that were its essence, or ideal plan, corresponding to something in the mind of the Creator. The science of comparative anatomy developed as a means of searching out these ideal plans, or archetypes. The central concept of comparative anatomy is homology, a term introduced by Sir Richard Owen, first Director of The Natural History Museum in London. Homology is the name given to the relation between, for example, the human arm, the foreleg of a horse, the wing

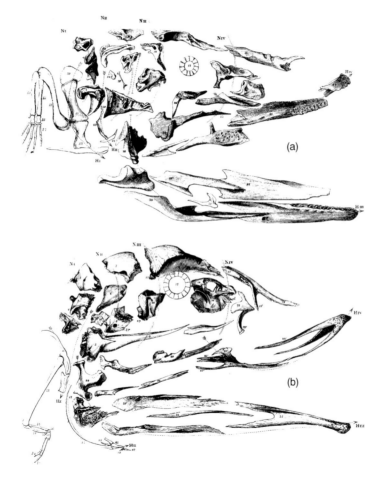

(a)

(b)

(c)

Figure 13.1

Homology

The skull and forelimb skeleton of a crocodile (a) and an ostrich (b) 'exploded' to show the separate bones, with homologous bones marked by numbers. By comparing the skeletons of many different vertebrates (fishes, reptiles, birds, mammals), Richard Owen deduced the imaginary animal shown in (c), his concept of the archetype or lowest common denominator of vertebrates. A modern evolutionist would construct a different imaginary animal, and call it the common ancestor of vertebrates.
From Owen's *Homologies of the Vertebrate Skeleton* (1848).

of a bat or bird and the flipper of a dolphin or turtle, or between our fingernails, the claws of a cat and the hooves of a horse or cow (Fig. 13.1). Homologous structures may differ in function or composition, but they correspond in relative position and, in the embryonic development of each individual, they arise from similar precursors. (In genetics homology is also used in a slightly different sense, for chromosomes or parts of chromosomes that correspond in structure and pair in meiosis; see page 14.)

By the time Darwin published *The Origin of Species*, Linnaean hierarchical classification and classical comparative anatomy – the search for the essence or archetype of each group by identifying homologous structures – were highly developed. Darwin was the first to suggest, by detailed argument, the significance of the natural hierarchy: that the relationship between the species of a genus, or the members of a family, is 'blood' relationship, caused by descent and divergence from a common ancestor. As shown in Fig. 13.2, the Linnaean classification can be interpreted

as resembling a family tree or genealogy, and the Darwinian interpretation is that a classification based on similarities and differences between species might accurately reflect the relationships of common ancestry among them. As Darwin wrote, 'our classifications will come to be, as far as they can be so made, genealogies; and will then truly give what may be called the plan of creation'.

When the theory of evolution became accepted, comparative anatomy received a new impetus, for the unravelling of homologies could now be seen as a key to understanding the course of evolution and 'archetypes' could be replaced by 'common ancestors'. A series of variations in some homologous structure, such as the vertebrate forelimb or skull, leads to the idea of the transformation of the ancestral (or primitive) structure into a variety of divergent, more advanced, specialized or derived conditions, each characteristic of a group of species. But Darwin's expectation that this would result in genealogical classifications has not yet been fully realized. This is mainly because

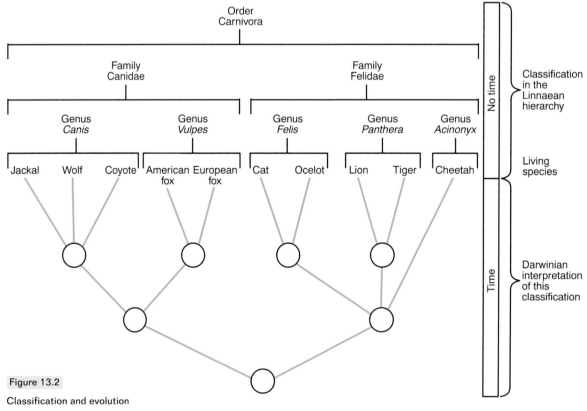

Figure 13.2

Classification and evolution

The relationships between the species in the Darwinian interpretation are due to descent from ancestral species (open circles), more or less remote in time.

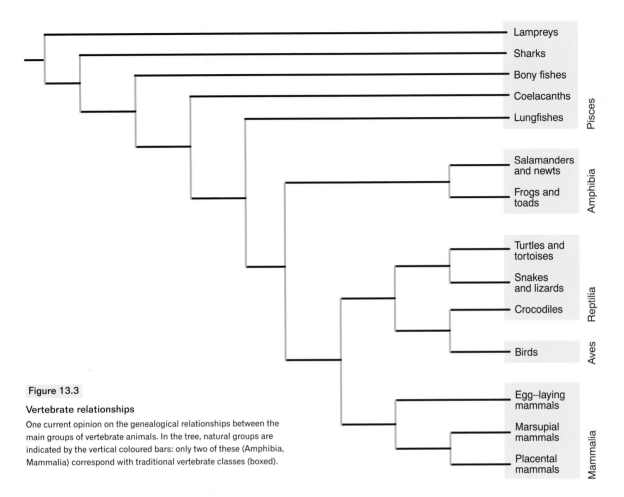

Figure 13.3

Vertebrate relationships

One current opinion on the genealogical relationships between the main groups of vertebrate animals. In the tree, natural groups are indicated by the vertical coloured bars: only two of these (Amphibia, Mammalia) correspond with traditional vertebrate classes (boxed).

of the enormity of the task, but is also because today, as in Linnaeus' time, classifications have two purposes – to express evolutionary relationships and to act as aides-mémoire or simple summaries of knowledge. These two aims come into conflict because relationships of common ancestry are almost invariably more complicated than the simple hierarchy that is conventional in Linnaean classification (see Fig. 13.3).

In evolutionary biology the concept of 'similarity' or 'resemblance' has three components.

- One is superficial, non-homologous resemblance (for example, the body form of a fish, a whale, a penguin and a seal); the Darwinian explanation for similarities like this is convergent evolution, by adaptation to similar ways of life.
- The second is resemblance due to retention of unchanged ancestral, primitive homologies (such as the

five toes of a frog, a lizard, a hedgehog and a human, whereas horses have one toe, deer and cattle two, rhinoceroses three, birds and dinosaurs four, snakes and whales none).

- Third, there is similarity in advanced or derived homologous features (like loss of the tail in apes and humans).

In traditional classification, similarity thought to be due to convergence is disregarded but the other two types of similarity are used. In genealogical classification, resemblance in primitive, ancestral features is disregarded. The example just mentioned – the five toes of humans, hedgehogs, lizards and frogs – tells us that all these animals are related by common ancestry and belong to a group comprising four-footed, five-toed vertebrates. At that level, a five-toed foot is an advanced feature compared with the fins of fishes. But if we want to know whether a lizard or a hedgehog is more closely related to humans or to frogs the

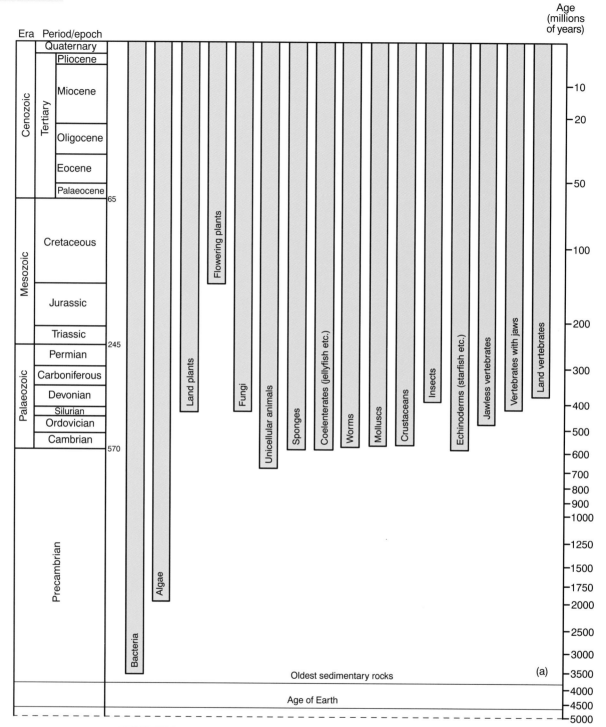

Figure 13.4

Geological time-scale and a summary of the fossil record

The time scale in these diagrams is logarithmic, giving greater space to more recent times. (b) (facing page) summarizes the time range of about 2500 subgroups (families) of organisms. The curve is based on the earliest fossil occurrence of each. About half extend back into the Mesozoic (65 million years or more), but only a handful are found in the Precambrian.

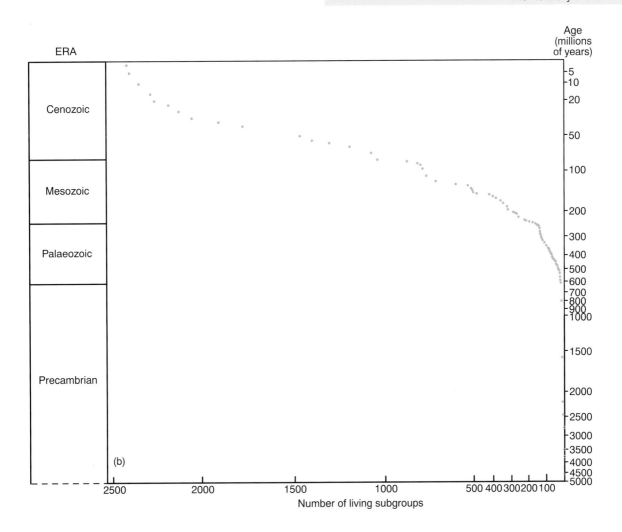

ERA

Age (millions of years)

Cenozoic

Mesozoic

Palaeozoic

Precambrian

(b)

Number of living subgroups

fact that they all have a similar foot is no longer any help, and we have to look for other more restricted homologies.

Lizards, hedgehogs and humans develop from an embryo which becomes enclosed in a special membrane, the amnion. This is an advanced feature that allows reptiles, birds and mammals to breed on dry land, and differentiates them from frogs and fishes, whose eggs must be laid in water. Hedgehogs and humans are warm-blooded, replace their teeth only once and have hair and other homologous features that relate them and differentiate them from lizards. By procedures like this, searching for homologies that characterize different groups and subgroups, it is possible to work out the genealogical relationships of most plants and animals in some detail, and to arrive at a classification which implies splitting of ancestral species into daughter-species whose descendants are now groups that are each other's closest relatives.

These relationships often demonstrate that traditional groups such as fishes and reptiles, which appeared to be as 'natural' to early evolutionists as they did to Linnaeus, are 'unnatural' in a genealogical sense, because they include species whose relationships lie elsewhere. This is illustrated in Fig. 13.3, in which the traditional classification of the main vertebrate groups is contrasted with the much more complex classification dictated by relationships of common ancestry.

Whether or not biologists should follow Darwin's advice, and make their classifications genealogical, is still controversial. The decision rests on a choice between a traditional classification whose main virtue is that it is stable and easily memorized and a more complicated classification, changing with new discoveries, whose main virtue is that it is a scientific theory of relationships. However the choice is made, the fact remains that the basis for a genealogical classification exists, in the unravelling of homologies.

The relevance of this to the theory of evolution is that homologous features are found at every level of diversity, from those that unite just two subspecies of Galapagos reptiles to those that unite all vertebrates or all multicellular animals or all living organisms (like the universal system of coding genetic information in nucleic acids). The resulting hierarchy of life, a classification of all organisms, is a clear indication that the causes of divergence between Galapagos species and subspecies also operated in the remote past, causing divergence within species which were ancestral to different major groups such as lungfishes and land vertebrates, or sharks and other jawed vertebrates, or plants and animals.

Darwin paid special attention to what he called 'rudimentary organs' (now usually called 'vestigial organs'), structures like the teeth found in embryonic baleen whales (which are toothless as adults) or the small pelvis and thigh bone buried in the body wall of whales and some snakes. He noted all sorts of vestigial organs in humans: our rudimentary body hair, our wisdom teeth (which may fail to erupt through the gum or cause problems when they do), the coccyx (a vestige of a tail) and the vermiform appendix, a troublesome appendage of the gut. These vestigial organs are clearly homologous with essential working parts in related species – teeth and hindlimbs in relatives of whales; fur, functional third molar teeth, a prehensile tail and the caecum in relatives of humans. Darwin found these vestiges or rudiments particularly interesting because they are difficult or impossible to explain under the 'plan of creation' view that he was opposing, because they imply imperfection or uselessness, not the perfect match of form and function found in their homologues. Under Darwin's view – descent with modification by natural selection – vestigial organs are readily explained as structures that were functional in ancestral species but are now selected against. It can be argued that we know too little of the function and construction of organisms, and that a whale's pelvis or the hair on our legs really serves some purpose but there are vestiges at the molecular level which escape that criticism (Section 13.3).

13.2 Fossils and geology

Fossils are the remains of long-dead organisms preserved, by natural burial, in the rocks of the earth's crust. With rare exceptions, all that is preserved is the hard parts or skeleton. But where the skeleton is moulded to the various parts of the organism, as in land plants, crustaceans, insects and vertebrates, a fossil may yield a great deal of information about its original owner.

When animals and plants die their remains are normally eaten by predators or scavengers and are eventually broken down by micro-organisms. Fossilization – in which the remains escape these hazards and are buried by sediment – is therefore a rare occurrence and requires unusual conditions. The chance that a fossil, once formed, will be found is even more remote, for only a minute proportion of fossil-bearing rocks are accessible to us, others having been eroded away, buried deep beneath the continents or oceans by earth movements, or consumed at the margins of shrinking tectonic plates. Darwin devoted two chapters of *The Origin of Species* to fossils, but spent the whole of the first in saying how imperfect the geological record of life is. It seemed obvious to him that, if his theory of evolution is correct, fossils ought to provide incontrovertible proof of it because each geological stratum should contain links between the species of earlier and later strata and, if sufficient fossils were collected, it would be possible to arrange them in ancestor-descendent sequences and so build up a precise picture of the course of evolution. This was not so in Darwin's time and today, after many more decades of assiduous fossil collecting, the picture still has extensive gaps.

The other objection that Darwin wished to counter concerned the age of the earth. At that time many editions of the Bible carried a marginal chronology, calculated from the genealogies of the prophets by a seventeenth-century archbishop, which placed the creation of the world in 4004 BC. The gradual changes that Darwin envisaged as causing the present diversity of species would clearly require a period of time vastly greater than that. Darwin and other nineteenth-century scientists attempted to calculate the age of the earth from the thickness of sedimentary rocks, the salt content of the oceans or the heat loss of the sun. Darwin arrived at an estimate of 200–400 million years. He felt that even this vast period might be too short to account for evolution but pointed out that we can hardly comprehend the meaning of millions of years.

During the twentieth century reliable methods of estimating the age of rock samples became available with the discovery of radioactivity. Radioactive elements are unstable, and decay at a rate which is constant for each type of atom. If a newly formed mineral containing a radioactive element is embedded in rock the products of radioactive decay will also be trapped, and the proportions of the element and its decay products will give an approximate age for the rock. The elements commonly used are uranium and thorium (which decay to produce lead and helium),

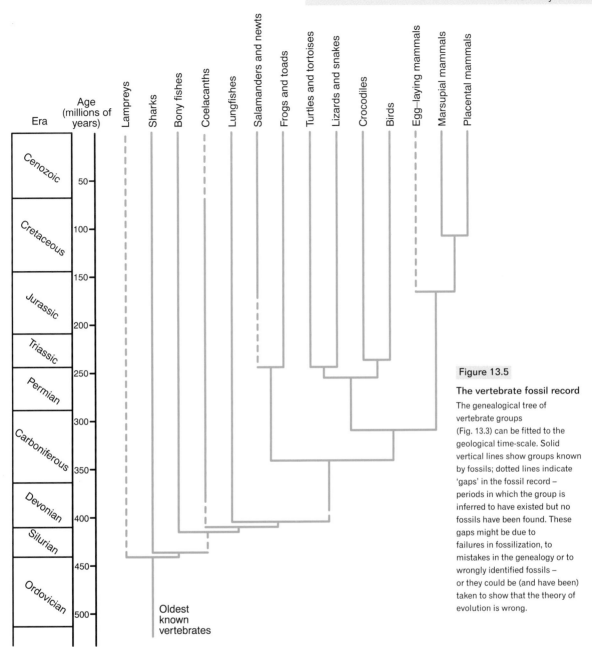

Figure 13.5

The vertebrate fossil record
The genealogical tree of vertebrate groups (Fig. 13.3) can be fitted to the geological time-scale. Solid vertical lines show groups known by fossils; dotted lines indicate 'gaps' in the fossil record – periods in which the group is inferred to have existed but no fossils have been found. These gaps might be due to failures in fossilization, to mistakes in the genealogy or to wrongly identified fossils – or they could be (and have been) taken to show that the theory of evolution is wrong.

radioactive potassium (producing argon), rubidium (producing strontium) and, in comparatively recent rocks (up to about 50 000 years old), radioactive carbon.

Real ages of particular rocks, as calculated from radioactive assays, can be combined with the traditional geological time-scale, which is derived from the fact that strata can be correlated by the fossils they contain* and that younger strata lie on top of older strata, to give a time-scale of earth history (Fig. 13.4(a)). This enormous period of time is surely

sufficient to meet any objections, and we can come back to the question of whether fossils, within this time-scale, provide a picture consonant with evolution. In broad terms they do. Figure 13.4(b) shows the range in time of the main groups of organisms and the proportion of subgroups recognizable at different times. The curve for the subgroups

*That different strata contain different fossils is evidence of change in the past, but is not direct evidence of evolution, or of the causes of evolution. Opponents of evolution have used fossils to support their own theories.

shows a regular decrease with increasing age, and in the main groups there is an orderly progression: the simplest in organization, like bacteria and simple seaweeds, appear before more highly organized organisms such as fungi and worms, and these in turn appear before flowering plants or land vertebrates. Thus the fossil record demonstrates progression in geological time, whether progression is defined as the development and further modification of homologous features or as increase in information content of DNA.

The fossil record is also consonant with evolution in an ecological sense – plants appear before animals (animals must depend on plants for food); land plants appear before land animals; plants with insect-pollinated flowers appear after insects, and so on.

There is a third way in which fossils support the theory of evolution. We saw in the last section that living species fall into a hierarchical classification, and that evolution offers an explanation – the hierarchy of groups within groups reflects genealogy and divergence from ancestral species in the more or less remote past. If the explanation is correct fossils should fit into the same system. Fossils from later (more recent) geological epochs should be members of living groups of low rank, and fossils from remote periods should fit only into groups of high rank. This is indeed so – for example, the genealogy of living vertebrate groups shown in Fig. 13.3 can be filled out by the addition of fossils, as in Fig. 13.5, the fossils providing a real time-scale for the various branching points and a check on the accuracy of the genealogy. As would be expected, we find that advanced groups such as birds appear later than groups like turtles and amphibians, which in turn appear later than the various types of fishes. Within each group we find the same sort of confirmation: no living mammal family is recognizable before the late Cretaceous, no living mammal order before the early Cretaceous – and as Jurassic mammals are followed back in time they become less diverse and less abundant.

Finally there is a point made by Darwin, that *all* fossils fit into the same hierarchy as living species: although we find fossil remains of many bizarre or unexpected extinct groups, none is entirely novel and all can be fitted into the hierarchy of living species at some level.

In several animal and plant groups enough fossils are known to bridge the wide gaps between existing types. In mammals, for example, the gap between horses, asses and zebras (genus *Equus*) and their closest living relatives, the rhinoceroses and tapirs, is filled by an extensive series of

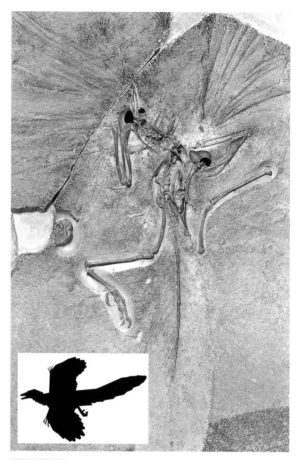

Figure 13.6

The earliest true bird, *Archaeopteryx lithographica*

This fossil comes from Upper Jurassic rocks, about 150 million years old, in Bavaria. Seven fossil skeletons have now been found, of which this one, discovered in 1861 and now in The Natural History Museum in London, was the first. *Archaeopteryx* is a bird because it has feathers, which have left impressions in the rock, and a wishbone. But it resembles dinosaurian reptiles, and differs from all living birds, in having teeth in the jaws, three clawed fingers on the wing, and a long bony tail with many vertebrae (birds have a rudimentary tail, the 'parson's nose'). About 1/4 natural size. The silhouette (inset) suggests how Archaeopteryx may have looked in flight.

fossils extending back 60 million years to a small animal, *Hyracotherium*, which can be distinguished from the rhinoceros–tapir group only by one or two horse-like details of the skull. There are many other examples of fossil 'missing links', such as *Archaeopteryx*, the Jurassic bird which links birds with dinosaurs (Fig. 13.6), or *Ichthyostega* and *Acanthostega* (Fig 13.7), late Devonian amphibians which link land vertebrates and the extinct choanate (having internal

Figure 13.7

**An early tetrapod,
*Acanthostega***

Tetrapods, or land vertebrates,
have four legs ending in feet
with digits (fingers, toes).
Primitive living tetrapods have
five fingers and toes, and until
recently this number was
thought to be ancestral for the
group. *Acanthostega*, from the
late Devonian of Greenland
(about 365 million years old), has
eight fingers and seven toes. It
also had fish-like gills and a fish-
like tail. All this suggests that
tetrapods first evolved in water
and that the 'conquest of the
land', with five fingers and toes,
came later.
(Courtesy of Mike Coates)

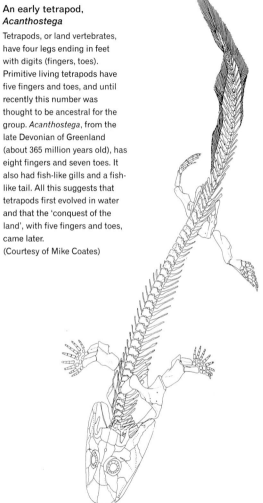

nostrils) fishes. But there are still great gaps in the fossil
record. Most of the major groups of animals (phyla) appear
fully fledged in early Cambrian rocks (Fig. 13.4), and we
know of no fossil forms linking them. Many different
'explanations' have been proposed – that the animals were
soft-bodied and would not be fossilized; that they lived in
places where fossil-bearing rocks were not formed; that the
early stages in the evolution of major groups are passed
through very rapidly in small populations. Perhaps the simplest explanation is that in many cases we do not yet know
what to look for, or how to recognize it if we found it.
Fossils may tell us many things, but one thing they can never
disclose is whether they were ancestors of anything else.

13.3 Molecular evolution

In the last 30 years, following the discovery that the messages in the genetic code concern protein structure, and the
development of methods for sequencing the amino acids of
protein chains and the letters of the DNA code, a revolutionary and powerful new means of investigating evolution
has appeared. We have already come across some of its
results, and have seen how family trees, like those produced
by comparing organisms (e.g. Figs 13.2 and 13.3), can be
reconstructed by comparing molecules (e.g. Figs 9.1 and
10.3). Homology, the relation between parts of organisms
that is explained by common ancestry, also exists at the
molecular level, in proteins and in DNA (Fig. 13.8). The
Darwinian explanation of homology (no other explanation
has been offered) is that organisms and their parts are related by common ancestry, by descent with modification from
pre-existing species that have split or speciated. By comparing homologous parts we can build up family trees, like
those in Figs 13.2 and 13.3, with species or groups of
species at the tips of the branches. In the same way, by comparing homologous genes or stretches of DNA we can build
up family trees like those in Figs 9.1, 10.3 and 10.6. But in
these molecular trees the genes or bits of DNA at the tips
of different branches may come from the same organism;
for example, in the tree of haemoglobin genes in Fig. 10.3
humans appear in five different places, and in the tree of *Alu*
sequences in Fig. 10.6 humans appear in 12 different places.
Gene trees may have different genes from the same species
at their tips but the explanation must be the same as for
trees of species – that genes, like species, are related by
common ancestry, by descent with modification from pre-
existing genes that have split or duplicated. Like species,
genes fall into a hierarchy of families and subfamilies
(Section 10.1). No one has yet tried to produce a Linnaean
classification of genes, but it is now common to find
references to classes and superfamilies of genes as well as to
families and subfamilies.

We have already discussed several members of one of
these families of genes, the globins (Section 10.3). The
protein products of globin genes contain about 150 amino
acids and include myoglobin (Fig. 4.5), which acts as an
oxygen store in the muscles of vertebrates, giving
mammalian muscle and the 'dark meat' of birds its red
coloration, the various haemoglobins (alpha, beta, gamma,
delta, etc.) which transport oxygen in the blood of mammals and other vertebrates, and more distantly related

```
                    1111111111222222222222333333333333444444444445555555555566666  ↓
          1234567890123456789012345678901234567890123456789012345678901234  ↓
                          Leu Leu Val Val Tyr Pro Trp Thr Gln Arg Phe Phe Asp
Human         TGACAAGAACA–GTTAGAG–TGTCCGAGGACCAACAGATGGGTACCTGGGTCTCCAAGAAACTG
Orang utan    TCACGAGAACA–GTTAGAG–TGTCCGAGGACCAACAGATGGGTACCTGGGTCTCCAAGAAACTG   2
Rhesus monkey TGACGAGAACAAGTTAGAG–TGTCCGAGGACCAACAGATGGGTACCTGGGTCTCCAAGAAACTG   2
Lemur         TAACGATAACAGGATAGAG–TATCTGAGGACCAACAGATGGGCACCTGGGTCTCCAAGAAACTG   8
Rabbit        TGGTGATAACAAGACAGAGATATCCGAGGACCAGCAGATAGGAACCTGGGTCTCTAAGAAGCTA  15
```

Figure 13.8

Homology at the DNA level

Part of the DNA sequence for a gamma haemoglobin gene in five species. The sequences can be aligned so that there is identity at the great majority of positions, and a few positions show variation (differences from the human sequence are shown in colour). For example, the adenine at position 5 in the human is homologous with a guanine in the others. On the right are the number of differences between each sample and the human sequence. This fragment is from the middle of the gene; part of the coding sequence begins at position 26, and the coded amino-acids are shown above the human sequence. Most of the differences between the sequences occur in the non-coding portion. They include inferred deletion or insertion of single bases, where a dash (a gap) is necessary to maintain identity between the sequences. The only differences in the coding portion are in the most distantly related animals, the lemur and rabbit; all are at third positions of triplets and do not alter the amino acid coded (compare the code in Fig. 4.9).

proteins such as the globins found in some invertebrates and plants.

In humans members of the globin family are found on three chromosomes. The gene for myoglobin is on chromosome 22 in the standard set (Fig. 4.3), the alpha haemoglobin subfamily of genes (containing, in sequence, zeta and a zeta pseudogene, two alpha pseudogenes, two alphas and theta) is on chromosome 16 and the beta subfamily

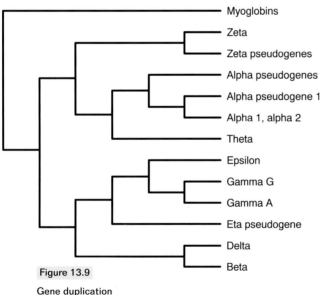

Figure 13.9

Gene duplication

A tree of the globin genes in humans.

(Fig. 10.1) is on chromosome 11. A tree can be built up by comparing the sequences of these human genes (Fig. 13.9); every event (split) in it must have occurred by duplication of an ancestral gene. If we add more species to the tree (Fig. 13.10) it will contain two kinds of events – splits between ancestral species (speciation) and splits between ancestral genes (duplications). The tips of the globin gene trees in Figs 13.9 and 13.10 are peppered with pseudogenes, dead or non-functional homologues – they have been called 'rusting hulks' – of functional genes. Pseudogenes have the same status as the 'rudimentary' or 'vestigial organs' that intrigued Darwin (Section 13.1). Whereas vestigial organs like human body hair and a whale's pelvis might have some function and so be explicable on the 'plan of creation' view that Darwin was opposing, pseudogenes, with mutations that prevent their translation and destroy their message, can be explained only as accidents of history, relics of common ancestry that have not yet decayed beyond recognition.

Some DNA sequence families consist of little but these 'rusting hulks' or accidents of history. As described in Section 10.3, there are about a million *Alu* sequences and 100 000 LINE1 sequences in every human cell nucleus, and together they account for almost 10% of our DNA. Like the members of the globin gene family, all *Alu* sequences are homologous and related by descent to (as yet) unknown functional progenitors, as are all LINE1 sequences. Another family of repeated sequences, widespread in animals and in plants, is in part homologous to the transfer RNA (Fig. 4.9)

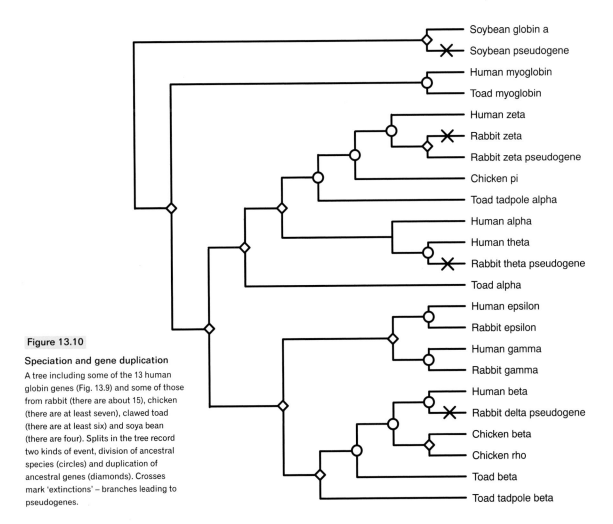

Figure 13.10

Speciation and gene duplication
A tree including some of the 13 human globin genes (Fig. 13.9) and some of those from rabbit (there are about 15), chicken (there are at least seven), clawed toad (there are at least six) and soya bean (there are four). Splits in the tree record two kinds of event, division of ancestral species (circles) and duplication of ancestral genes (diamonds). Crosses mark 'extinctions' – branches leading to pseudogenes.

for the amino acid lysine, copied back into the DNA again and again. At the molecular level the equivalent of Darwin's vestigial organs – 'junk DNA' – takes up a lot of space.

In 1863, Darwin's friend T.H. Huxley published his first book, *Evidence as to Man's Place in Nature*, within a couple of years of his first reading of *The Origin of Species*. It is worth concentrating on that small part of the hierarchy of life covered by the title of Huxley's book to see what molecular data has to say about the relationships of our own species. The close similarity between apes and humans was recognized by Linnaeus, who was the first to classify them together. But to take the next step and infer that this similarity indicates genealogical relationship required such boldness or disregard for sensibilities that Darwin side-stepped the issue in *The Origin of Species*; in the whole volume his only reference to human evolution is one cryptic sentence: 'Light will be thrown on the origin of man and his history.' He tackled the subject properly in *The Descent of Man*, published in 1871, where he wrote 'the gorilla and chimpanzee … are now man's nearest allies.' Today it is accepted that the apes are our closest living relatives, and that among the four types of living ape (gibbons, orang utans, gorillas and chimpanzees) gorillas and chimpanzees are closest to us. Nevertheless, the difference between us and these apes seems so great that we tend to assume that the relationship is not too close. In terms of classification, T.H. Huxley placed humans on their own in a suborder of primates; Darwin thought 'that this rank is too high, and

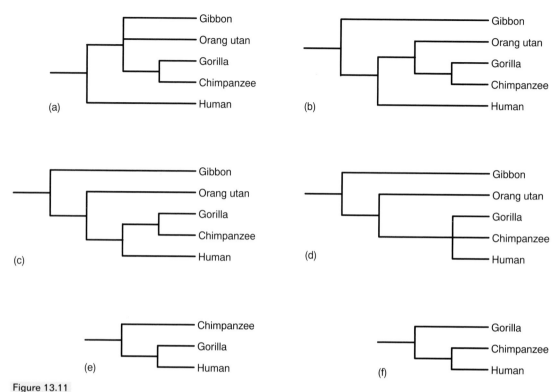

Figure 13.11

How are we related to apes?

In (a), the traditional, pre-Darwinian view, apes form one group and humans are distinct from them.
(b) A scheme still found in some books, with chimpanzees, gorillas and orang utans as our closest
relatives and gibbons more distant. (c) The theory found in Darwin's *Descent of Man* (1871) and
many modern text-books puts chimpanzees and gorillas as our closest relatives.
(d) A compromise, dictated mainly by studies of proteins and chromosomes in the 1970s and 1980s
implies that we cannot yet decide whether we are closest to gorillas (e) or chimpanzees (f). The
current theory, strongly supported by the most recent evidence of long sequences of DNA, is (f).

that man ought to form merely a Family, or possibly even
only a Sub-family'; yet T.H. Huxley's grandson, Sir Julian
Huxley, proposed in the 1940s that humans might deserve a
kingdom to themselves – Psychozoa.

In terms of history, Darwin thought that humans
diverged from the apes somewhere between Eocene and
late Miocene times (a span of time now dated at about
10–50 million years ago; Fig. 13.4), and by the 1960s, with
the help of fossils found since Darwin's time, the date of
human divergence was narrowed down to early Miocene or
late Oligocene times (20–30 million years ago). Molecular
biology began to impinge on these ideas in the late 1960s,
when molecular clocks (see page 63) were first proposed.

When applied to comparisons between proteins of humans
and apes the clock indicated that humans diverged from
apes no more than five million years ago. This proposal,
which seemed absurd at first, is now agreed to be not far
from the truth, and older fossils once placed on the human
lineage are now seen to belong elsewhere.

At the amino acid and DNA sequence level many molecules and genes bearing on human relationships have now
been investigated.

- Cytochrome *c*, a respiratory protein, has exactly the
  same amino acid sequence in humans and chimpanzees,
  and this sequence differs from that of the rhesus monkey
  by one mutation.

- Fibrinopeptides A and B (blood plasma proteins) are identical in humans, chimpanzees and gorillas and differ from orang utans by two mutations and from gibbons by three or four.
- Human alpha and beta haemoglobin chains are the same as those of chimpanzees, while gorillas differ from humans in one amino acid in each chain and gibbons differ in three.
- Human delta haemoglobin differs from that of chimpanzees and gorillas (which are the same) in one mutation, and of gibbons in two.
- Human myoglobin differs from that of chimpanzee, gorilla and gibbon by one mutation (a different one in each case), and from orang utan by two.
- In the enzyme carbonic anhydrase humans and chimpanzees differ by two mutations, and humans and orang utans differ by nine.

At the DNA sequence level, impressive evidence comes from the 17 000 bases of the beta haemoglobin gene cluster (mentioned in Section 11.3; part of the alignment is shown in Fig. 13.8) that has been sequenced in two humans, pygmy and common chimpanzee, gorilla, orang utan, gibbon and various other primates. Comparison of the various sequences shows that humans and chimpanzees share 25 inferred mutations whereas the nearest competitor, a pairing of chimpanzees with gorilla, is supported only by seven inferred mutations. On that evidence, it is about 99.9% certain that chimpanzees are our closest relatives. Comparison of another huge set of sequences, the entire DNA of the mitochondria (see page 142) – about 15 000 bases – in three humans and all four ape lineages (chimp, gorilla, orang, gibbon), also puts chimpanzees as our closest relatives, with confidence greater than 99.9%.

The meaning of these comparisons is summarized in Fig. 13.11, showing the evolution of our understanding of the relationship between humans and apes. The general conclusion from the molecules that have so far been investigated is that the genetic differences between a chimpanzee and a gorilla, minor to our eyes, are actually slightly greater than those between a human and a chimpanzee, while the difference between an orang utan and a chimpanzee, perhaps even slighter to us, is far greater than that between chimpanzee and human. In genetic terms we are hardly more distinct from chimpanzees than are subspecies in other groups of animals (Fig. 11.4). This conclusion is something of a shock to our views on the uniqueness of humans, and is discussed further in Section 16.1.

13.4 'Hopeful monsters'

So far in this chapter we have found that several lines of investigation, more or less independent of one another, indicate that large-scale differences like those between human and mouse or elephant and oak tree have the same sort of causes as those which differentiate closely related species. Humans and mice, elephants and oak trees, are related by common ancestry and owe their differences to divergence, by accumulation of mutations, over enormous periods of time. The evidence for this is homology, the fact that these organisms share homologous features at many levels: molecular (DNA, RNA, homologous proteins, etc.), structural (cellular organization, etc.), physiological (pathways of oxygen metabolism, etc.), in the life cycle (diploid and haploid phases, etc.) and so on.

Yet are we justified, on this evidence, in making the leap from small-scale changes like selection in peppered moths, speciation in the Galapagos and duplication or divergence of globin genes, to large-scale results such as the existence of elephants and oak trees? Some evolutionists have felt unhappy about this: while accepting that the gradual accumulation of small differences, due to point mutations, gene duplications and chromosome mutations, will account for divergence within such groups as butterflies or birds, they have felt that the original appearance of birds, or land vertebrates, or vertebrates as a whole requires innovations that cannot be satisfactorily explained by gradual, small-scale changes. So they have supposed that major innovations arise at one step, by large-scale, favourable mutations, or **macromutations**. The useful name 'hopeful monsters' is given to the original lucky carriers of such mutations. This name refers to the difference between the potential of the carrier of a typical 'monstrous' or harmful large-scale mutation and the carrier of a true innovation. For example, we can contrast a mutation causing shortening of the tail in a cat with that in the reptile-like bird *Archaeopteryx* (Fig. 13.6). In the first case the result is a Manx cat, a curiosity. In the second case the feathers on the long tail of *Archaeopteryx* might become concentrated into the fan with which modern birds control their flight.

The main reason for inventing these macromutations is that some features of plants and animals can hardly be imagined as arising by gradual steps; the adaptive value of the perfected structure is easily seen but intermediate steps seem to be useless, or even harmful. For example, what use is a lens in the eye unless it works? A distorting lens might be worse than no lens at all. What use are feathers unless they are 'proper' feathers? What use is a lung that is half-developed, and cannot give you enough oxygen? How can the segmentation of an animal like an earthworm or a centipede arise bit by bit? An animal is either segmented or it is not.

The usual answer to such questions is that they are due only to failure of imagination. Rudimentary feathers would be useful to an ancestral bird if, like living birds, it was warm-blooded, for they would conserve heat by insulating it. Half a lung, or a quarter of a lung, would be useful to a fish, for the air-bubble would increase its buoyancy so that it would have to expend less energy in staying off the bottom. These two examples illustrate the principle of **preadaptation**, which explains puzzles like feathers and lungs by showing that intermediate stages in their evolution could be promoted by selection not because of their present use (flying, breathing), but for quite a different reason (heat conservation, buoyancy). We can imagine selection working on such structures for one function up to a certain threshold, beyond which a new function becomes possible, so that adaptive change is diverted into a wholly new path.

Some of the innovations requiring 'hopeful monsters' have yielded to explanations of this sort, but others remain unsolved and the idea of macromutations as a force in evolution persists. What form might these macromutations take? One class of macromutations, the doubling of the entire genetic complement of an organism in polyploidy, has already been discussed (Section 6.3). This is certainly an evolutionary force in plants, and does happen rarely in animals. Yet here the difficulty is one outlined in Section 6.3. The individual in which the number of chromosomes doubles is effectively sterilized. Meiosis is seriously disturbed, and if gametes are produced they can meet compatible gametes only by self-fertilization, or in the almost impossibly unlikely event of meeting gametes from an individual of the opposite sex that has also suffered chromosome doubling.

Gene duplications of the kind that produced the different globin genes (Fig. 13.9), and which must be responsible for many biochemical innovations, are a less conspicuous type of macromutation. Here the problems of infertility are less because if the duplicated chromosome section is small it may hardly affect meiosis. Yet if the effects of the mutation are large, the problem of spreading the mutation is still there, for who will breed with a monster, hopeful or otherwise?

Today speculation on macromutations mainly concerns their effects on regulatory genes, the genes that switch on and off batteries of protein-producing genes. We still know little about these regulatory master genes. Their main field of operation must be embryonic development, in which cells are marshalled and differentiated to form the various organs. The best candidates to date are the **homeobox** family of genes. The spectacular effects of homeobox gene mutations were first seen in *Drosophila*, early in the history of genetics. Carriers of some of these mutations certainly qualify as monsters – though without much hope (Fig. 13.12). The general patterns in **homeosis** (homeobox mutation) are for a part or organ to be replaced by another that is normally found elsewhere, or to be duplicated. For example, the eyes may be replaced by a pair of wings or the antennae by legs, and legs or the thorax may be duplicated. In the last few years we have learned that the genes responsible for these effects are part of a family of genes that is very widely distributed, occurring in all multicellular animals and in at least some plants and fungi. In *Drosophila*, mutations of *Antennapaedia* (*Antp*) genes include replacement of the antennae by legs (Fig. 13.12), and *Antp* genes have now been found in animals as diverse as corals and humans. In mice and humans the *Antp* subfamily of homeobox genes contains about 40 members, arranged in

Figure 13.12

A homeotic mutation in *Drosophila*
A leg has developed in place of the antenna.
SEM Dr R. Turner.
(Courtsey of Flybase at Harvard University)

four clusters, each on a different chromosome. These four clusters are homologous with a single group of about ten genes in *Drosophila* and with a group of three in a coral; they are clearly the result of a history of duplication and divergence like that of globin genes (Fig. 13.9).

Homeobox genes are active in early development of the embryo, and have two functions:

- They encode a short regulatory protein which binds to a particular sequence of bases in DNA and either enhances or represses gene expression.
- They specify proteins which are expressed in complex patterns that determine the basic geometry of the organism.

We can imagine that point mutations in genes like these, controlling early development, might produce changes that are small initially, when they act in the embryo, but large in the adult through amplification in the course of development. Point mutations of this sort need not impair fertility, yet if their effects are large the problem of the monster finding a mate remains.

One possibility is inbreeding amongst the offspring of the individual that first develops the mutation in its germ cells. On average, half of the offspring will carry the mutation, and through inbreeding they could establish it. In this form the hopeful monster theory is probable, although some orthodox evolutionists still reject it.

14 Proof and disproof; science and politics

The wrong view of science betrays itself in the craving to be right.
Karl Popper

I have steadily endeavoured to keep my mind free, so as to give up any hypothesis, however much beloved (and I cannot resist forming one on every subject) as soon as facts are shown to be opposed to it.
Charles Darwin, *Autobiography*

The first edition of this book contained a short chapter with the heading 'Proof and disproof,' applying my (naive) understanding of some of the ideas current in philosophy of science of the 1970s to evolutionary theory. The chapter was troublesome in various ways, but in rewriting the book I eventually decided to leave it much as it was, because I still think the ideas are interesting, and to add a commentary ('Science and politics').

14.1 Science versus pseudo-science

The great tragedy of Science – the slaying of a beautiful hypothesis by an ugly fact
T.H. Huxley

Is the theory of evolution by natural selection proved? After so many pages of fact and argument, some may be disconcerted by a negative answer and to read that certainty can no more be found in science than in any other way of thought. These ideas come from Sir Karl Popper (1902–1994), the great philosopher of science. Popper argued that proof, or certainty, exists only in mathematics and in logic, where it is trivial in the sense that the proven conclusions were already hidden in the premises. For him, science is distinguished from non-science (not nonsense), or metaphysics, or myth, not by proof but by the possibility of disproof. The only characteristic of scientific theories is that they have consequences which might be falsified by observation or experiment, and a scientist is a person who is willing to relinquish his theory when it is falsified or refuted. Pseudo-scientific or metaphysical theories do not expose themselves to test or disproof in this way.

The classic example in science is Newton's theory of gravitation, which was the foundation of physics for more than 200 years, and seemed to be the epitome of established knowledge, or scientific certainty. Yet in this century Newton's theory was replaced by Einstein's, and this replacement was the consequence of observations (principally of the eclipse of the sun in 1919) which tested the two theories and disproved one of them. Of course, Newton's theory was not shown to be completely wrong; it was found to be less universal and so further from the truth than Einstein's.

Nor was Einstein's theory shown to be true; it may be replaced by another more inclusive or general theory, even closer to the truth, at any time. And we shall never know whether this process of replacement has stopped, for that would mean that we had arrived at the truth, and even if we have found it we have no criteria for recognizing truth (sincere belief and consensus are often mistaken for such criteria).

Popper's favourite examples of pseudo-scientific or metaphysical theories are psychoanalysis and astrology. Freudian psychology, for example, predicts that neuroses and other mental disorders are the result of incidents in early childhood.

But there is no conceivable observation that can clash with this theory, and any behaviour or recollected incident may seem to confirm a diagnosis of a particular patient (once a person is authoritatively diagnosed as mentally ill, any behaviour seems to fit the diagnosis). The difference between a scientist and a pseudo-scientist is, in Popper's view, that the first will look for the most severe tests of his theories and will not take evasive action if they fail those tests, while the pseudo-scientist will look for evidence confirming his ideas and, if he feels his theory is threatened, may avoid refutation by erecting subsidiary, defensive theories around it.

14.2   Is evolution science?

If we accept Popper's distinctions between science and non-science we must ask first whether the theory of evolution by natural selection is scientific or pseudo-scientific (metaphysical). That question covers two quite separate aspects of evolutionary theory. The first is the general thesis that evolution has occurred – all animal and plant species are related by common ancestry – and the second is a special theory of mechanism, that the cause of evolution is natural selection (in fact Darwin accepted the first idea a couple of years before he thought of the second).

The first, general, theory (that evolution has occurred) explains the history of life as a single process of species-splitting and progression. That process must be unique and unrepeatable, like the history of England. Before Darwin, species were generally thought to be fixed and immutable, each with some discoverable and universal essence, like the elements or chemical compounds. Darwin explained species as temporary, local things, each with a beginning and an end depending on contingencies of history. He converted biology from a study of universals, like chemistry, to a study of individuals, like history. So the general theory of evolution is a historical theory, about unique events – and unique events are, by some definitions, not part of science for they are unrepeatable and so not subject to test. Historians cannot predict the future (or are deluded when they try) and they cannot explain the past, but only interpret it. And there is usually no decisive way of testing their alternative interpretations. For the same reasons, evolutionary biologists can make few predictions about the future evolution of any particular species, and they cannot explain past evolution but produce only interpretations or stories about it. Yet biologists have enormous advantages over historians. First, they have a coherent, and scientific, theory of genetics, and their interpretations must be consistent with it. Second they have one basic tool, homology (Section 13.1). And third they have the universal scientific principle of parsimony, or economy of hypothesis, also known as Occam's razor: the simplest story is the best. Despite these advantages for the evolutionist it remains true that there are no laws of evolution comparable to the laws of physics, just as there are no laws of history.

The general theory of evolution is thus neither fully scientific (like physics, for example) nor unscientific (like history). Although it has no laws it does have rules, and it does make general predictions about the properties of organisms. It therefore lays itself open to disproof. Darwin cited several sorts of observations which would, in his view, destroy his theory. In this he was certainly more candid than his opponents. The potential tests Darwin mentioned are:

- 'If it could be demonstrated that any complex organ existed, which could not possibly have been formed by numerous, successive, slight modifications, my theory would absolutely break down';
- 'certain naturalists believe that very many structures have been created for beauty in the eyes of man, or for mere variety. This doctrine, if true, would be absolutely fatal to my theory';
- 'if it could be proved that any part of the structure of any one species had been formed for the exclusive good of another species, it would annihilate my theory'.

Darwin's potential tests may strike the reader as pretty feeble, or as tests of natural selection rather than evolution, but many discoveries, not foreseen by Darwin, provide more severe tests of the theory. These include Mendelian genetics, the real age of the earth, the universality of DNA and the genetic code and (most recently and spectacularly) the evidence from DNA sequences of innumerable 'vestigial organs' at the molecular level. Evolution has survived all of these with flying colours. Darwin could not possibly have predicted that the hereditary material (of which he knew nothing) would turn out to be littered with rubbish, with 'rusting hulks' like the delta haemoglobin pseudogene found in Old World monkeys (page 69), or with meaningless repeated sequences like the shared *Alu* sequences in apes and humans (Fig. 10.6). An interesting argument is that in the law courts (where proof 'beyond reasonable doubt' is required), cases of plagiarism or breach of copyright will be settled in the plaintiff's favour if it can be shown that the text (or whatever) is supposed to have been copied contains errors present in the original. Similarly, in tracing the texts of ancient authors, the best evidence that two versions are copies one from another or from the same original is when both contain the same errors. A charming example is an intrusive colon within a phrase in two fourteenth-century texts of Euripides: one colon turned out to be a scrap of straw embedded in the paper, proving that the other text was a later copy. Shared pseudogenes, or shared *Alu* sequences, may have the same significance – like shared misprints they can have come about only by shared descent.

Turning now to the second or special theory of mechanism, many critics have held that natural selection, as the cause of evolution, is not scientific because the expression 'survival of the fittest' makes no predictions except 'what survives is fit', and so is tautologous, or an empty repetition of words. For example, one answer to the question 'which are the fittest?' might be 'those that survive', so that 'survival of the fittest' means only 'survival of the survivors'. Indeed, natural selection theory can be presented in the form of a deductive argument, for example:

- All organisms must reproduce.
- All organisms exhibit hereditary variations.
- Hereditary variations differ in their effect on reproduction.
- Therefore variations with favourable effects on reproduction will succeed, those with unfavourable effects will fail, and organisms will change.

In this sense, natural selection is not a scientific theory but a truism, something that is proven to be true, like one of Euclid's theorems: if statements 1–3 are true, so is statement 4. This argument shows that natural selection must occur but it does not say that natural selection is the only cause of evolution, and when natural selection is generalized as the explanation of all evolutionary change or of every feature of every organism it becomes so all-embracing that it is in much the same class as Freudian psychology and astrology. An additional difficulty is the one explained in Section 7.2, in the discussion of genetic drift. Modern evolutionary theory does not say that all evolutionary change is caused by natural selection: random effects, like genetic drift, have played a very important part at the level of DNA (Section 9.3). Natural selection is therefore protected from falsification by the alternative explanation – random or neutral change.

14.3 Alternative theories

Using Popper's criterion we must conclude that evolutionary theory is not testable in the same way as a theory in physics, chemistry or genetics, by experiments designed to falsify it. But the essence of scientific method is not to test a single theory to destruction; it is to test two (or more) rival theories, like Newton's and Einstein's, and to accept the one that passes more or stricter tests until a better theory turns up. We must look at evolution theory and natural selection theory in terms of performance against the competition.

The belief that all organisms are related by descent and have diverged through a natural, historical process has only one main competitor, creation theory, though there are different stories of how the Creator went about His work. All creation theories are purely metaphysical. They make no predictions about the activities of the Creator, except that life as we know it is the result of His plan. Since we do not know the plan, no observation can be inconsistent with it. At one extreme there is the fundamentalist view that apparent evidence for evolution, such as fossils, was built into the newly created rocks to tempt us or test our faith. At the other extreme is the person to whom evidence of evolution only pushes the activity of the Creator further and further into the past. Both these modifications of the original creation myths are typical evasive moves, avoiding refutation or confrontation by modifying the original theory, or erecting subsidiary defensive theories around it.

Natural selection theory – the belief that evolution has been caused by the gradual accumulation, in different diverging lineages, of small, favourable genetic mutations – has several competitors. First are the vitalist theories, which accept that evolution has occurred but propose that its course has been guided or directed towards certain ends by some vital force, universal consciousness or striving for perfection. Second is the idea that evolution proceeds by the inheritance of modifications acquired, through use or disuse, during the life of individual organisms, the theory first proposed (in 1809) by the French naturalist Lamarck. Third are a number of theories that accept evolution by natural selection on the small scale but invoke large-scale, unique mutations as the source of major steps in evolution like the origin of the animal phyla, or of birds or flowering plants (Section 13.4). And fourth, the neutral theory of molecular evolution (Sections 9.2 and 9.3) proposes that, at the DNA level, much evolutionary change is caused not by selection of favourable mutations but through fixation, by chance, of mutations that are neutral (or nearly so) in their effects and so invisible to natural selection.

How does natural selection theory compare with these? Vitalist theories are purely metaphysical, for they make no predictions about the past or future activities of the various occult agents which are supposed to guide or direct evolution. Most vitalist theories seem to be only modified creation theories.

Lamarckian theory, hinging on the inheritance of acquired characters (education by the environment rather than selection by it), suffered a severe reverse early in this century from the work of geneticists, who could find no good examples of the process and no mechanism that could bring it about. The theory seems to have been effectively disproved by the unravelling of the genetic code and the way in which it is transcribed and translated, for this is effectively a one-way process with no feedback through which new information can be passed from the cell into the

DNA (information can pass from the cell into the DNA, for example in the form of processed pseudogenes – see page 117 – but the information so passed seems to be nonsense).

Macromutation theories (Section 13.4) suppose that major groups and innovations appear not by gradual selective change but by a special class of large-scale, unique mutations. Because these theories invoke unique events in the distant past they are not testable. They predict that 'missing links' will continue to elude us, but even finding some such links will not disprove the theories, for there will still be other groups that are not linked. I know of one attempt to test macromutation theory, a demonstration that the adult size of members of species in many groups of animals does not vary gradually but in jumps, the ratio between the size of one species and another being 1:2, or 1:4 or 1:8. In primates (humans, apes and monkeys) for example, the ratios are 1:8:64:512, rising in eightfold steps. These results are consistent with macromutation, and are interesting, but they were produced by believers in macromutation, whose statistical methods have already been criticized by mathematical selectionists.

Finally, there is the neutral theory of molecular evolution (Sections 9.2 and 9.3), with its message that those changes in DNA that are least subject to natural selection occur most rapidly, with mutation pressure rather than selection pressure as the driving force. This theory was first proposed in the late 1960s when very little was known about molecular evolution; it was based on calculations of the theoretical rate of fixation of neutral mutations, and on estimates of rates of change in mammalian haemoglobins and other proteins, calculated from the few amino acid sequences then available. It was a bold proposal, not only because it opposed the panselectionism then in favour (Section 10.4) but also because it was made in mathematical terms and had obvious testable consequences, such as the existence of molecular clocks (Section 9.1), and that silent changes (mostly in the third position of the DNA triplets coding amino acids) should outnumber changes (mostly in the first and second positions) that replace the coded amino acid (Section 9.2). That last prediction has been met (see Fig. 9.2), and other subsequent discoveries match the predictions of neutral theory, such as 'genes-in-pieces' and pseudogenes, both pseudogenes and the untranslated introns of genes evolving much faster than coding sequences (see Fig. 9.2). Neutral theory is therefore more like a theory in physics or chemistry than other evolutionary theories; it makes bold, testable predictions, and

can be put in a law-like form: changes in DNA that are less likely to be subject to natural selection occur more rapidly. This law has not yet been proved false.

14.4 A metaphysical research programme?

Darwin ended his Introduction to *The Origin of Species* with the words 'I am convinced that Natural Selection has been the main but not exclusive means of modification' (in the Fifth Edition (1869) he altered this to ' … has been the most important, but not the exclusive means …'). By 'modification' he meant change in the form or structure of organisms, change in their phenotypes. Natural selection is the only mechanism yet proposed that will explain adaptation, and at the level of the phenotype we have no reason to doubt his statement. In the genotype, the DNA or hereditary material, things are different because the dominant 'means of modification' seems to be neutral change, evolution despite natural selection rather than because of it. But, recent discoveries about the prevalence of neutral change in DNA that is invisible to natural selection do nothing to explain the existence of whales and elephants, fleas and butterflies or ferns and oak trees. It is impossible to believe that the different phenotypes of these organisms are merely the result of neutral change in DNA, although the importance of neutral change at the molecular level should temper the ultra-Darwinian or panselectionist view that every feature of every organism is shaped by natural selection. At present we are left with neo-Darwinian theory: evolution has occurred and has been directed mainly by natural selection of phenotypes, with major contributions from neutral drift in the genotype, and perhaps the occasional hopeful monster. In this form the theory is not 'scientific' by Popper's standards: indeed, Popper has called the theory of evolution 'a metaphysical research programme'. He meant that, though the theory is closer to metaphysics than are theories in physics, accepting it as true gives us a research programme, a new way of looking at and investigating the world. And through this research programme we can make progress in understanding the world. As one criterion of progress in science Popper offers this: 'If the progress is significant then the new problems will differ from the old problems: the new problems will be on a radically different level of depth.' It is surely true that the problems occupying today's workers in molecular evolution are on a radically different level of depth from those which interested mid-Victorian evolutionists.

Yet Popper warns of a danger: 'A theory, even a scientific theory, may became an intellectual fashion, a substitute for religion, an entrenched dogma.' This has certainly been true of evolutionary theory, and it leads into a different view of scientific progress that has been developed mainly by the American philosopher and historian, Thomas S. Kuhn. His ideas are as much concerned with the sociology of science as with its philosophy – they deal with how science *is* done, not, like Popper's, with how it *might be* done.

In Kuhn's view the progress of science is neither orderly nor rational but proceeds in lurches, with long calm periods and occasional revolutions. In the calm periods – what Kuhn calls 'normal science' – some high-level theory is accepted by the scientific community and research is devoted to solving puzzles within the framework provided by that theory. Observations that conflict with the theory (that seem to disprove it) are pushed to one side or avoided by defensive subtheories. But sooner or later, perhaps because of a build-up of these awkward observations, a crisis develops and a new high-level theory is proposed. Some scientists are 'converted' to the new theory and begin puzzle-solving within it. Others will stick with the old theory, and a dialogue between the two groups, or a decisive confrontation of the two theories, is hardly possible, for members of opposing groups interpret the world in the light of their own theory so that words have different meanings for them. In these conflicts the eventual success of the new (or the old) theory is due not to proof, disproof or logic but to factors that are effective in politics, religion or art – conversion, faith, taste.

Kuhn's views, cynical as they may sound when summarized like this, certainly find echoes in the history of evolutionary theory. As a simple example of the impossibility of dialogue between supporters of opposing theories, I will cite Darwin's most powerful opponent in America, the great Swiss-born naturalist Louis Agassiz. He was unmoved by *The Origin of Species*, and remained true to the theory that species were individually created and immutable. He saw Darwin's view that species may change and grade into one another as a denial of the reality of species, and put to Darwin what he saw as an unanswerable argument: 'If species do not exist, how can they vary? And if individuals alone exist, how can differences among them prove the variability of species?' To Darwin, Agassiz's unanswerable conundrum was 'an absurd logical quibble'. Both were correct within their own framework of thought, and neither could see his opponent's point. In the same way, to Agassiz

the word 'evolution' meant the unfolding of the individual's potential, in development from the egg, and he regarded the new use of the word by Darwin and his followers as a gross misuse. Such complaints and misunderstandings are common in all arguments. Here, Agassiz lost the day, and the Darwinian revolution triumphed. Following it, we can recognize a series of subsidiary revolutions: the Mendelian revolution at the beginning of this century (Chapter 4), the population genetics revolution in the 1930s (Chapter 7), the DNA revolution in the 1950s, and the neutralist revolution (Chapter 9), which began in the late 1960s and is still a field of conflict. Each of these has introduced a new and deeper set of problems and a new vocabulary.

No doubt other revolutions are in store, and whether we choose to follow Popper's or Kuhn's understanding of science the one lesson we can learn from both of these thinkers is that today's theory of evolution is unlikely to be the whole truth. It is essential to keep in mind the distinction between the general theory – evolution has occurred and species are related by descent – and theories of mechanism – natural selection, neutralism, etc. Today's theory, accepting that evolution has occurred and explaining it by neo-Darwinism plus neutralism, is the best that we have. It is a fruitful theory, a stimulus to thought and research, and we should accept it until nature prompts someone to think of one that is better or more complete.

14.5   Science and politics

There is no alternative.
Margaret Thatcher

This chapter is so far much as it was in the first edition, though it then included less on neutral change in DNA and nothing on pseudogenes and other molecular misprints, which were then undiscovered. As a result it was perhaps a bit more agnostic than the version you have just read. What follows is a commentary on some events since.

I should have expected trouble from my attempt to dabble in philosophy of science, after one referee of the manuscript for the first edition wrote of this chapter 'stuff like this makes me want to reach for a gun.' Perhaps scepticism runs in my profession; my predecessor as curator of fossil fishes at The Natural History Museum, Errol White (1901–1985), when President of the Linnaean Society of London in the 1960s published a presidential address including 'I have often thought how little I should like to have to prove organic evolution in a court of law. In my experience of fossil fishes, while one can see the general drift of evolution readily enough, when it comes to pin-pointing the linkages ... the links are invariably either missing altogether or faulty ... We still do not know the mechanics of evolution ... and we shall certainly not advance matters by jumping up and down shrilling "Darwin is God and I, So-and-so, am his prophet".'

In 1981 The Natural History Museum celebrated the centenary of its opening in South Kensington after separation from the British Museum in Bloomsbury. As part of the centenary events a new permanent exhibit on the 'Origin of Species' was opened (to be conveniently in place for the commemoration of the centenary of Darwin's death in 1982). The new exhibit was opened by Sir Andrew Huxley, then President of the Royal Society and a Trustee of the Museum. Sir Andrew, a Nobel prizewinner in physiology, is the grandson of T.H. Huxley, 'Darwin's bulldog', and the half-brother of Aldous and Julian Huxley. His speech at the opening ceremony, and his Presidential Address to the Royal Society that year, emphasized a controversy that had occupied much space in *Nature* over the preceding couple of years. The controversy centred on the work of Natural History Museum scientists in classifying organisms (particularly the use of branching diagrams like Fig. 13.11) and on related aspects of the Museum's exhibition policy (such as the use of branching diagrams as aids in exhibits on dinosaurs and on human evolution). *Nature* published much correspondence around these topics and early in 1981 two leading articles under the titles 'Darwin's death in South Kensington' and 'How true is the theory of evolution?' The first was critical of the Museum's exhibition policy, and the second discussed responses from Museum scientists (denying that evolution is a proven fact) and brought in Sir Karl Popper's philosophy of science.

In opening the Museum's 'Origin of Species' exhibit, Huxley commented on these matters and assured his audience that 'Darwin is alive and well in South Kensington.' But on the following day *Nature* came out with another leading article – 'Does creation deserve equal time?'. For in March 1981 the state of Arkansas signed into law the 'Balanced treatment for creation-science and evolution-science Act', which began 'Public schools within this State shall give balanced treatment to creation-science and to evolution-science.' *Nature's* editorial linked the Arkansas law with the Museum's new exhibit, because at its entrance were two panels, one summarizing Darwin's explanation

for the diversity of species and the other saying that there were other explanations, including 'the theory of creation'. 'Tolerance has gone too far' said the editorial. The following week *Nature* carried a review of the Museum's new exhibit, mostly devoted to one corner of it, a mini-theatre in which a short film explained the current status of the theory of evolution. 'Scientists', said the reviewer, 'expect museums to represent their views as cogently as possible'. Without descending 'to the half-truths and double-talk of political propaganda' they should not blunt the edge of the message 'with the hair-splitting logic-chopping of the philosophy of science.'

The relevance of all this is that the film in the mini-theatre was based on the text of this chapter, as published in the first edition, converted into a dialogue between two actors. I agree with *Nature's* reviewer that the film gave an unnecessarily negative impression of neo-Darwinian theory. I felt that this impression had only one cause – the actor playing the sceptic came over as the stronger character. If the roles had been reversed the film might have got by without negative comment. So, here, theatre rather than science or philosophy seemed to resolve the question.

The film was soon withdrawn from the exhibit, but *Nature's* correspondence columns continued to be filled with letters on whether evolution is or is not metaphysics. Popper had contributed to this debate in 1980 with a letter to *New Scientist* that included 'some people think that I have denied scientific character to the historical sciences, such as ... the history of the evolution of life on Earth. ... This is a mistake ... historical sciences have in my opinion scientific character: their hypotheses can in many cases be *tested*. ... some people would think that the historical sciences are untestable because they describe unique events. However, the description of unique events can very often be tested by deriving from them testable predictions or retrodictions.' Popper gave no example of such a prediction, but one often quoted (attributed to J.B.S. Haldane) is that finding human bones in Carboniferous rocks (about 350 million years old) would falsify the general theory of evolution. However, faced with good evidence of such a discovery I feel that biologists would not discard evolution but protect their theory by finding some alternative explanation; for example, that a time machine will one day be invented. As many have argued, and as Popper has acknowledged, falsification can be as slippery as verification. Nevertheless, in 1978 Popper wrote that the Darwinian theory of common descent 'has been well tested' (he did not say how) and

'That the theory of natural selection may be so formulated that it is far from tautological. In this case it is not only testable, but it turns out to be not strictly universally true. … Yet in every particular case it is a challenging research programme to show how far natural selection can possibly be held responsible for the evolution of a particular organ or behavioural programme.'

In December 1981 the job that Errol White had not wanted ('to have to prove organic evolution in a court of law') seemed to become necessary. The Arkansas Balanced Treatment Act was challenged in the courts by a consortium of clergy, educational organizations and citizens, and a number of distinguished evolutionists converged on Little Rock to serve as witnesses in a trial before Judge William Overton. In the event, proving evolution turned out to be unnecessary because the alternative, 'creation science', was on trial. Judge Overton delivered a remarkably forthright and hard-hitting judgment in January 1982. He found that science has four essential features:

- It is guided by natural law, and is explanatory by reference to natural law ('natural' here is the antithesis of 'supernatural' agencies, such as the Creator's power).
- Science is testable against the empirical world, our world of experience.
- Its conclusions are tentative, not the final word.
- It is falsifiable.

The judge found that 'creation science' failed to meet all of these requirements, that its practitioners could report no real research, that teachers who had been required to prepare a curriculum for it could find nothing to teach and (to paraphrase) that, all in all, the opposition to the flourishing enterprise of evolutionary biology consisted of people who were committed to a literal reading of the first chapters of Genesis. Judge Overton therefore ruled that the Balanced Treatment Act violated the separation of church and state required by the constitution of the United States.

The Arkansas judgment applied only in that state, and during the 1980s other states considered or passed 'equal time' laws that required teachers in public schools, if they taught evolution, to give equal time or equal status to alternative explanations. In 1987, in a case arising from a law passed in Louisiana, the Supreme Court outlawed such proposals throughout the USA. Nevertheless, in the 1990s comparable acts have been considered by state or local legislature in Alaska, California, Texas and Vermont. The emphasis has now shifted from 'creation science' (ruled by the courts to be thinly veiled Genesis) to attempts requiring teachers to consider 'alternatives to evolutionary theory', 'other theories of the origin of man' or 'scientific evidence against evolution'. No doubt, if or when such legislation is passed and challenged in the courts the judge will find that the impetus behind it is the same as that behind the Arkansas law – 'a religious crusade' (as Judge Overton put it) 'motivated by adherence to Fundamentalist beliefs in the inerrancy of the book of Genesis.'

The creationist movement, strongest in the USA, is also highly visible elsewhere – in Australia, for example. In the UK creationism maintains a much lower profile, presumably because, with no written constitution, an established church and religion taught in the public schools, there is no point in a political campaign opposing teaching of evolution. Because creationists lack scientific research or evidence to support such theories as a young earth (10 000 years old), a world-wide flood (Noah's) and separate ancestry for humans and apes, their common tactic is to attack evolution by hunting out debate or dissent among evolutionary biologists. When I published the first edition of this book I was hardly aware of creationism but, during the 1980s, like many other biologists I learned that one should think carefully about candour in argument (in publications, lectures or correspondence) in case one was furnishing creationist campaigners with ammunition in the form of 'quotable quotes', often taken out of context. Biologists, particularly in North America, took creationism seriously enough in the 1980s to produce a string of books promoting evolution and showing the errors of creationist anti-science.

14.6 Summary

There is, I believe, no necessary conflict between evolution and religion, a topic I leave to Chapter 16. I will close by summing up my own opinion on the truth of evolution. In terms of mechanism, or causes of evolutionary change, the neutral theory of molecular evolution is a scientific theory; it can be put in law-like form: *changes in DNA that are less likely to be subject to natural selection occur more rapidly*. This law is tested every time homologous DNA sequences are compared, and explains observations (summarized in Chapters 9 and 10) that are otherwise inexplicable. But neutral theory assumes (or includes) truth of the general theory – common ancestry or Darwin's 'descent with modification' – and 'misprints' shared between species, like the pseudogenes or reversed *Alu* sequences (Fig. 10.6), are (to me) incontrovertible evidence of common descent. I see the general historical theory, common descent, as being as firmly established as just about anything else in history. We have compelling reasons to believe that Napoleon and the Roman empire existed, although we don't know every detail of what went on in Napoleon's life or in Rome and its colonies; it is much the same with evolution. There is abundant documentary evidence for Napoleon and the Roman empire; there is abundant evidence for common descent in the hierarchy of homologies at both the structural and morphological level, though those documents may not be so easy to read. Natural selection, Darwin's explanation for descent with modification, undoubtedly works in both its negative form (preventing change; explaining why some bits of DNA evolve more slowly than others or do not evolve at all, for example) and positive form (promoting change; explaining adaptation or why we always need new insecticides and antibiotics). The most difficult problems in evolutionary biology are in deciding or discovering the extent to which positive natural selection is responsible for the diversity of life.

15 The origin and early evolution of life

*Analogy would lead me one step further, namely, to the belief that all
animals and plants are descended from some one prototype.*
Charles Darwin

Up to this point we have been discussing the theory of evolution in general terms. The theory leads us to believe that all present-day animals and plants are related by community of descent. This implies that life arose only once, and this book would be incomplete without a summary of the current ideas on how life first arose and on the earliest stages in evolution. The topic is certainly intriguing but, since it concerns unique events in the remote past, it is far more speculative than evolution theory. Rather than qualify every statement by 'it may be', or 'perhaps', I will ask the reader to remember the comments on historical explanation in Section 14.2, and to understand that the historical parts of

this chapter are a plausible and consistent story, not fact.

Before discussing conditions in the early stages of earth's history, we need some idea of the simplest and most primitive organisms alive today.

15.1 Prokaryotes and eukaryotes

So far in this book I have written as if plants and animals were the only sorts of organisms. This is a simplification. Plants and animals were the only kingdoms recognized by Linnaeus, and that usage persisted until the middle of this century. But today five kingdoms are often accepted: Animalia, Plantae, Fungi, Protista and Monera. The scientific

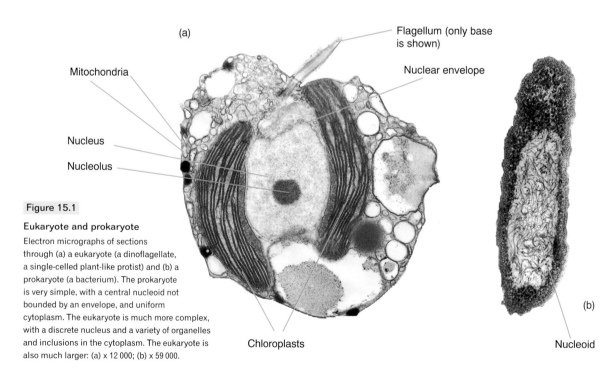

(a)

Flagellum (only base
is shown)

Mitochondria

Nuclear envelope

Nucleus

Nucleolus

Chloroplasts

Nucleoid

(b)

Figure 15.1

Eukaryote and prokaryote
Electron micrographs of sections
through (a) a eukaryote (a dinoflagellate,
a single-celled plant-like protist) and (b) a
prokaryote (a bacterium). The prokaryote
is very simple, with a central nucleoid not
bounded by an envelope, and uniform
cytoplasm. The eukaryote is much more complex,
with a discrete nucleus and a variety of organelles
and inclusions in the cytoplasm. The eukaryote is
also much larger: (a) x 12 000; (b) x 59 000.

meaning of the first three is close enough to common usage for them to be self-explanatory. **Protists** are single-celled organisms, or organisms that form multicellular colonies without differentiation or co-operation between the individual cells. Some, such as amoeba, are animal-like in mode of life, some are plant-like, with rigid cellulose cell walls and chlorophyll, and others are fungus-like. (Protists raise acutely the problem of traditional versus genealogical classification (Section 13.1), for in the evolutionary tree (see Fig. 15.3) they form an array of branches below animals, plants and fungi and in a genealogical classification each of these protist branches deserves at least the same kingdom rank as those three. A compromise is to classify protists in three kingdoms.) Many protists are mobile, driving themselves along by one or more long, whip-like structures (flagella), or by many small hairs (cilia). **Monerans** are 'microbes', bacteria and their relatives: minute, single-celled organisms usually less than 0.01 mm in diameter or length.

The four (or more) higher kingdoms – plants, animals, fungi and protists – are grouped together as **eukaryotes**; the monerans are **prokaryotes**. The prokaryote/eukaryote distinction (Fig. 15.1) is one of the most profound and important in biology; the major break in the hierarchy of life is not in an obvious place like between animals and plants but between two kinds of micro-organisms, mostly invisible to the naked eye.

- The cell nucleus, the chromosomes contained in the nucleus and the cycles of nuclear division (mitosis and meiosis) and of sexual reproduction described in the early parts of this book are all features of eukaryotes; none occurs in prokaryotes.
- Instead of a nucleus, prokaryotes have a 'nucleoid', an area of the cell that is not bounded by a nuclear envelope.
- Instead of a set of chromosomes, prokaryotes have a **chromoneme**, a double helix of DNA that is not insulated by protein like the DNA of eukaryotes and in which the two ends are usually joined to form a closed loop or ring.
- Instead of undergoing a mitotic cycle prokaryotes synthesize DNA continuously while active, and divide by simple fission.
- Sexual reproduction, in which two haploid gametes unite to form a diploid zygote, occurs only in eukaryotes; those prokaryotes that undergo a 'sexual' process simply pass chromonemal material from one cell, the donor, to another, the recipient.

- Prokaryotes do not have cilia or true flagella, and those which are motile move by gliding, by wriggling the whole cell, as in spirochaete bacteria, or by means of bacterial 'flagella' – inflexible helical rods that rotate like propellers.
- 'Genes-in-pieces' (see Fig. 4.9), with the coding parts of the gene (exons) separated by untranslated introns, are almost unknown in prokaryotes.

The other essential difference between prokaryotes and eukaryotes is that prokaryotes do not have **mitochondria** or **chloroplasts**. Virtually every eukaryote cell contains numerous mitochondria. There are minute inclusions, bounded by a membrane with internal partitions. They contain the enzyme systems responsible for oxygen metabolism, the oxidation or 'burning' of food molecules. Chloroplasts are the green, red, yellow or brown bodies found in the cells of plants). Like mitochondria, they are membrane-bounded, have internal partitions, and there are usually several to each cell. Chloroplasts contain, and synthesize, chlorophylls, and are the sites of photosynthesis, the process in which energy from sunlight is 'fixed', sugars are synthesized from carbon dioxide and water, and oxygen is given off as a waste product. In prokaryotes the enzymes responsible for oxygen metabolism are scattered throughout the cell, or are missing altogether. Prokaryotes lacking these enzymes live in **anaerobic** environments (without oxygen), and many of them die if they come into contact with oxygen. Photosynthesis occurs in some prokaryotes, but again the pigments and enzymes responsible are not packaged. And in many prokaryotes photosynthesis follows chemical pathways that do not produce oxygen.

These differences between prokaryotes and eukaryotes are profound: they are greater than those between a human and a tree. In almost every feature prokaryotes are simpler than eukaryotes, and this suggests that in speculating about the origin of life, it is the origin of prokaryotes that we should think of, not the more complicated eukaryotes. The primacy of prokaryotes is confirmed by the fossil record. Figure 13.4 shows the geological time-scale and the first appearance of fossil representatives of different groups. Recognizable multicellular animals and plants do not appear until latest Precambrian times, about 600–700 million years ago. Protists, unicellular eukaryotes, may have existed almost 2000 million years ago, but the prokaryotes extend back about 3500 million years. The earth is some 4600 million years old, so for about one-third of that enormous span of time the only fossils found are prokaryotes, resembling living bacteria. The problem of the origin of life can therefore be narrowed down to the problem of how prokaryotes originated.

The relationship between prokaryotes and eukaryotes is discussed later in this chapter – at this stage it is necessary to say only that prokaryotes and eukaryotes have so many fundamental features in common, especially the method of transmitting information in a triplet code in DNA and translating it into proteins through RNA, that the most economical theory is that eukaryotes evolved from prokaryotes and that life as we know it originated once only, when these genetic mechanisms became established. One other important fact bears on this conclusion – the 'handedness' or dissymmetry of biological molecules.

Sugars, amino acids and many other chemical compounds are asymmetric molecules, which can exist in two forms, mirror images of each other (Fig. 15.2). These two forms can be detected by their effect on polarized light. A solution containing equal quantities of the two forms – 'left-handed' and 'right-handed' molecules – will transmit a beam of polarized light unchanged but a solution containing mainly or only right-handed molecules will rotate the beam in one direction, and one of mostly left-handed molecules will rotate it the other way. According to their effect on the beam the two varieties are called laevo-rotatory (to the left, abbreviated as L) and dextro-rotatory (to the right, D). The handedness of a simple molecule does not affect its chemical properties, except in reactions with other asymmetric molecules.

If a sugar or an amino acid is synthesized in the laboratory the product will be a mixture of L and D molecules, in roughly equal numbers, but in living organisms of every sort all the amino acids are L-amino acids and in nucleic acids all the sugars are the D form. It is presumably accidental that life originated with this combination because the

(a)

Figure 15.2

Dissymmetry of biological molecules

(a) The two snail shells, with clockwise and anticlockwise spirals, are an example of mirror-image structures.
(b) Models of the molecule of the amino acid alanine.
The L (laevo-rotatory) form is the only one synthesized in living organisms.

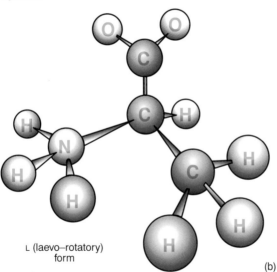

L (laevo–rotatory) form

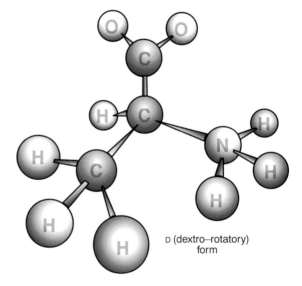

D (dextro–rotatory) form

(b)

alternative, D-amino acids and L sugars, should work just as well. This uniformity in handedness is further evidence that life originated once only.

15.2 Diversity of prokaryotes: rooting the tree of life

We know that prokaryotes, the simplest surviving forms of life, first appeared on earth at some time between its origin about 4600 million years ago and their first occurrence as fossils, about 3500 million years ago. We can only speculate about how they originated, but there are four interconnected ways of approaching the problem.

- We can survey living prokaryotes and try to reconstruct their common ancestor, the simplest conceivable prokaryote.
- By comparing duplicated genes we may be able to reach back beyond that ancestor, to estimate some of the earliest components of the genetic machinery.
- We can try to reconstruct the conditions that existed on the earth's surface in those remote times.
- We can simulate those conditions experimentally and see what is produced.

The eukaryote/prokaryote distinction was not clearly recognized until the middle of the twentieth century. About

Figure 15.3

The tree of life
A tree including examples of the three main divisions of organisms (eukaryotes, Archaea and Bacteria). Placing a root in the tree is problematic.

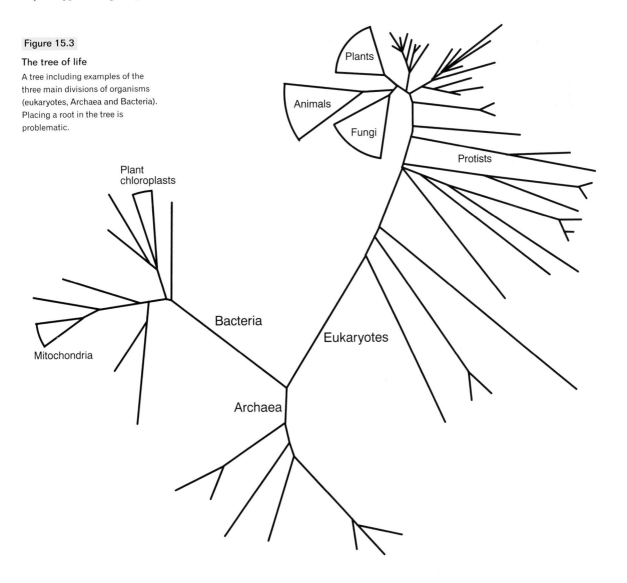

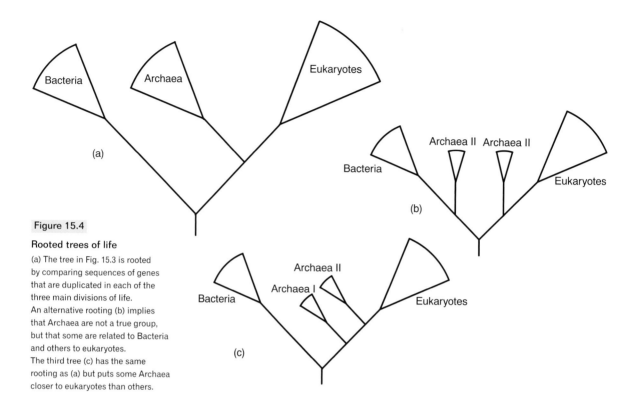

Figure 15.4

Rooted trees of life

(a) The tree in Fig. 15.3 is rooted by comparing sequences of genes that are duplicated in each of the three main divisions of life.
An alternative rooting (b) implies that Archaea are not a true group, but that some are related to Bacteria and others to eukaryotes.
The third tree (c) has the same rooting as (a) but puts some Archaea closer to eukaryotes than others.

1980, as methods of sequencing DNA developed, it became clear that there are two major kinds of prokaryotes, now placed in two separate kingdoms – Eubacteria (or just Bacteria) and Archaebacteria (or Archaea). In the rest of this chapter Bacteria (with a capital) implies the restricted sense of Eubacteria/Bacteria, and bacteria means prokaryotes in general. Since prokaryotes are less than 0.01 mm long (and many of them less than 0.001 mm), it is not surprising that prokaryote classification depends mainly on DNA sequencing and other molecular techniques rather than on comparing structures visible under the microscope. Bacteria include most of the organisms that we understand by bacteria – for example, the cause of nasty smells in the kitchen and of the success of sewage works, the infective agents in cholera and diphtheria, our intestinal bacterium *Escherichia coli* and the photosynthetic blue–green 'algae'. The Archaea (Archaebacteria) are less numerous or diverse than Bacteria and are found only in extreme environments, uninhabitable to most other organisms, where they survive by strange metabolisms. The concept 'food' almost loses meaning when applied to the bizarre diets of these prokaryotes: some 'feed' on hydrogen or sulphur, or on simple inorganic compounds like hydrogen sulphide (here, 'feed' means 'carry out controlled chemical reactions and utilize the energy released'). One of the two main groups of Archaea, the thermophiles ('heat-lovers'), live only at extremely high temperatures, 80°C or above, in places like volcanic hot springs and the recently discovered deep-sea hydrothermal vents. The other group, the halophiles ('salt-lovers'), live only in extremely high salt concentrations, such as in the Dead Sea, salt pans and subterranean brines associated with oil reservoirs.

The differences between Archaea and Bacteria are subtle, and at the chemical or molecular level. Interestingly, when their different features are compared with corresponding features in eukaryotes, Archaea seem closer to eukaryotes than are Bacteria. Trees made by comparing homologous DNA sequences from the three groups (Fig. 15.3) are generally shown in radiating form, because of difficulties in deciding where to place the root of the tree. In most evolutionary trees (like Figs 9.1, 10.3 and 13.3) the root is placed by reference to other, more distantly related, organisms but here the tree covers the most basic divisions of life, and 'more distantly related' organisms do not exist.

However, there is a strategy that may allow glimpses into a more distant past, and so help to root the tree – to compare genes that duplicated below that root and so are represented by divergent descendants in each of the main branches (the duplicated globin genes in Figs 13.9 and 13.10 are a similar example within the eukaryote branch).

Comparisons of this sort, using three different pairs of duplicated genes, agree that the root lies somewhere on the segment separating Bacteria from Archaea plus eukaryotes (Fig. 15.4(a)), but other evidence conflicts with this and the question is not yet settled – reconstructing events that took place billions of years ago is bound to have its problems. An interesting alternative view is that the root lies within Archaea (Fig. 15.4(b)), and there are strong indications (from DNA sequences) that Archaea is not a natural group because some thermophiles seem more closely related to eukaryotes than are the halophiles and other thermophiles (Fig. 15.4(c)). Further, in molecular trees of Bacteria the two groups closest to the root are also thermophiles, living in hot, anaerobic marine sediments or subterranean brines. All of this implies that the earliest prokaryotes were found in extremely hot environments without oxygen and lived by metabolizing sulphur, hydrogen or simple inorganic compounds.

Some of these basic prokaryotes are so small (only about 0.001 mm in diameter) and have so little DNA that they have information and room for only a few hundred different enzymes. Nevertheless, they have, as far as we know, the complete machinery for replication and protein synthesis – DNA with a triplet code, the three types of RNA (messenger, transfer, ribosomal) and the enzymes controlling this system. Such creatures are very close to the root of the tree of life, but are not necessarily to the simplest imaginable organism.

15.3 The RNA world

In some warm little pond ...
Charles Darwin

As we have just seen, the simplest living organisms use the same basic machinery as our own cells, storing information in DNA and transcribing and translating it into protein with the help of RNA. Imagining how this system originated leads to a 'chicken-and-egg' conundrum: which came first – DNA or proteins – given that proteins, the end product, are essential as enzymes in translating and copying the

information in DNA? A possible solution comes from discoveries, first made in the early 1980s, that some forms of RNA are 'self-splicing' and can act as catalysts, promoting steps in reactions involved in self-replication and in protein formation. Speculation on the origin of life is now centred on an 'RNA world' that may have preceded and developed into the DNA/RNA/protein world of prokaryotes and eukaryotes.

Fossils very similar to living prokaryotes are found in rocks about 3500 million years old. The earth originated about 4600 million years ago, but for the first 400–500 million years was an inhospitable place, bombarded by meteorites and too hot for liquid water, which is essential for RNA chemistry. So the RNA world, if it existed, could have developed and given rise to the DNA/protein world only in a (relatively) brief period, perhaps from about 4100 to 3800 million years ago. For some eminent scientists, including Francis Crick, the chances of life originating in such a period are too slim; they prefer the theory that life arose elsewhere and that earth was 'seeded' by primitive organisms arriving in meteorites or other extraterrestrial material. But this theory just transfers the problem of the origin of life to some other planet, and we might as well speculate about conditions here, where they are more accessible.

At the time of the birth and development of the RNA world, terrestrial conditions were very different from today. Many primitive prokaryotes can live only in the absence of oxygen. Oxygen is a reactive element, easily combined with others, and free oxygen could hardly have existed in the atmosphere of the primitive earth. It seems certain that the free oxygen (which now forms about 20% of our atmosphere) has been produced by life, as a by-product of past photosynthetic activity in plants and blue–green Bacteria. Absence of oxygen in the primitive atmosphere means that the first organisms would have been anaerobic, but has another consequence. Today, the earth is surrounded by a layer of ozone ('condensed' oxygen) in the upper atmosphere. Recent concern about destruction or erosion of this ozone layer, apparently by chemicals (such as chlorofluorocarbons) generated by human activity, centres on the fact that ozone absorbs ultraviolet light, and shields the earth's surface from this important part of the sun's energy. Ultraviolet radiation promotes mutation (increasing skin cancers, for example), and strong ultraviolet radiation kills all forms of life. So, in the primitive atmosphere with no oxygen there was no ozone shield against ultraviolet radiation. That atmosphere was formed primarily by out-gassing from the earth's crust and mantle, as occurs in volcanic activity today, and would have consisted mainly of water vapour, carbon dioxide, nitrogen and sulphur dioxide. Methane, ammonia and carbon monoxide might also have been present. Liquid water was also certainly present, in the primitive oceans, because life depends on water. The ocean would have contained all the atmospheric gases in solution.

With no ozone shield against ultraviolet radiation, the energy from the sun reaching the earth's surface would have been greater than it is today. The radioactivity of the earth's crust would also have been greater. Other sources of energy would include electric discharge (e.g. from lightning) and volcanic activity. Experiments have been conducted in which gas mixtures simulating various estimates of the primitive atmosphere are subjected to ultraviolet radiation, or electric discharge, heat, shock waves and so on. The products of these experiments are surprising. A variety of organic compounds is produced and major components are amino acids (including most of the biologically important ones), purines and pyrimidines (including the four bases in RNA – adenine, cytosine, guanine and uracil – but not thymine, which replaces uracil in DNA), sugars, porphyrins (molecules which are the forerunners of important biological compounds like vitamin B12, chlorophylls, etc.) and complex tar-like compounds which defy analysis. All these products found have equal quantities of L and D molecules. That such molecules do form in abiotic (lifeless) conditions is confirmed by the presence of the same compounds in meteorites, of amino acids in rock samples from the moon and, most recently, by detecting the spectral signature of the amino acid glycine in a cloud of star-forming gas in a distant part of the galaxy. In the primitive oceans these and others products would accumulate over long periods of time, producing a thin 'soup' of organic molecules, perhaps with slicks (or even surface layers) of oils and tars, which would shield the underlying water from ultraviolet radiation.

Local concentrations of organic compounds might build up in this 'primitive soup' in a variety of ways. In shallow water this might happen by evaporation of lakes or ponds, by freezing (a very efficient mechanism, since water is gradually withdrawn from solution as ice) or by adsorption on the surfaces of solids such as clay particles or on the surface film of the water. An alternative scenario is suggested in deeper waters by recent discoveries in the deep sea of thriving colonies of primitive prokaryotes in hydrothermal vents, where heat and sulphur compounds are abundant, and where the overlying water shields organisms from ultraviolet radiation.

In the RNA world the minimal organism is visualized as an RNA molecule that is capable of self-replication, acting as a template on which a copy is synthesized. There is still a wide gap between such a simple system and an unorganized mixture of the simple molecules (with sugars in both L and D forms) from which it is built up. So far it has proved difficult to find plausible reactions that will combine purines and pyrimidines with ribose sugar and phosphates to produce nucleotides, the building blocks of nucleic acids. But once a self-replicating system appears, mistakes in replication (mutations) will occur and will be subject to natural selection. It is still a long step from replicating nucleic acids, subject to natural selection, to the unique and complex co-operative system of proteins (built up from L amino acids) and nucleic acids that characterizes life as we know it. Here speculation centres on whether DNA, the information store, was invented before proteins, or whether proteins, the catalysts, came first.

Once primeval organisms appeared, whether by a unique series of chance reactions or by some inevitable process that we have not yet thought of, they would not have been short of food, because to the first organisms the 'primeval soup' would be a nutritious broth of energy-rich molecules. Reproduction in this fertile medium might be extremely rapid, and the 'soup' would soon become thinner, as the molecules that had accumulated before the origin of life were used up. Natural selection would then begin in earnest, for any mutant would have a great advantage if it could synthesize from simpler, abundant molecules some essential chemical that was in short supply. Presumably it was by successive mutational steps like this that the first metabolic pathways were built up.

In these early organisms, the most essential pathways would be those concerned with reproduction, with improving accuracy and efficiency in replicating nucleic acids. Energy for these processes must be generated by the organism, and possible sources might include simple fermentation processes; the ability to 'fix' carbon dioxide, producing organic compounds by reactions involving atmospheric hydrogen; the ability to 'fix' nitrogen, producing nitrates and organic nitrogen compounds; by the synthesis of porphyrins and related compounds which are forerunners of photosynthetic pigments (the chlorophylls) and of cytochromes (the basis of oxygen metabolism). The photosynthesizers would come first, and would originally have been anaerobic, using carbon dioxide and hydrogen sulphide, hydrogen or ammonia, as do some living bacteria. The aerobic (oxygen-producing) photosynthesis of blue–green Bacteria uses carbon dioxide and water and is a more complicated process, requiring greater energy to split the water molecule. Because this sort of photosynthesis produces oxygen, no organism could evolve it unless it already had some means of dealing with oxygen. So some form of oxygen metabolism, using cytochromes, must have preceded aerobic photosynthesis. Once the blue–green Bacteria appeared oxygen would begin to accumulate in the atmosphere. This produced a crisis, because many existing organisms would be anaerobes, unable to tolerate free oxygen. The anaerobes would be killed off or restricted to oxygen-free environments unless they could develop means of tolerating or (better) utilizing oxygen. Various prokaryotes tolerate oxygen in low concentrations, but others have come to use it and rely on it. Some perform simple oxidation reactions (of iron, nitrogen, sulphides, etc.) as a source of energy; others use cytochrome and a complex enzyme pathway to 'breathe' oxygen in the same way as we do.

15.4 The origin of eukaryotes

Simple eukaryotes, resembling living unicellular algae, are first confirmed in the fossil record about 1500 million years ago and first suspected in rocks almost 2000 million years old. Early eukaryote fossils are recognized by a crude criterion – cell size; prokaryotes are generally smaller than about 0.01 mm and eukaryotes are generally larger. But some living eukaryotes are much smaller (down to 0.001 mm), and some of the older fossils could well be eukaryotes. Trees based on molecular comparisons (see Fig. 15.4) imply that eukaryotes may be as old as Archaea. Oxygen, produced by photosynthesis in blue–green Bacteria, must have begun to accumulate in the atmosphere by 2500 million years ago as 'red beds', sedimentary rocks containing oxidized iron (rust), are dated from then onwards. This atmospheric oxygen would initiate the ozone shield against lethal solar ultraviolet radiation.

Eukaryotes are almost all aerobic – they require and metabolize oxygen by an enzyme pathway (the citric acid cycle) which includes cytochrome. These reactions take place in packets within the cell, the mitochondria, which are present in most eukaryotic cells. A remarkable feature of mitochondria is that they have their own genetic system, distinct from the chromosomal genes in the cell nucleus.

Mitochondria contain DNA and RNA, and synthesize their own enzymes. They are capable of division (reproduction) and lead a semi-autonomous life. When a eukaryote cell divides, each of the daughter cells receives mitochondria in the cytoplasm. In some eukaryotes, division of the mitochondria is synchronized with nuclear division; in others, mitochondria become aligned on the spindle in mitosis and are segregated into two groups at the same time as the chromosomes. However, mitochondria are not completely autonomous, for some of their functions are coded and controlled by nuclear genes. Mitochondrial DNA is very different from that in the nucleus and is more like the DNA of prokaryotes. These strange facts can all be explained by one proposal – that mitochondria were originally independent, free-living prokaryotic organisms resembling aerobic bacteria. Their inclusion in eukaryotic cells is an example of **symbiosis**.

Symbiosis is an association between two (or more) different organisms that is of benefit to both (or all) partners (symbionts). Symbiotic associations range from the casual or part-time, like birds that pick ticks off cattle, to those that are essential for the survival of the partners, such as the obligatory association between termites and the wood-digesting protists that inhabit the termite gut. Perhaps the most familiar example of symbiosis is the lichens, a group of 'plants', each of which is a mixed colony of a fungus and an alga, able to survive and grow in more rigorous environments than either partner could tolerate alone. There are many symbiotic relationships in which one partner lives inside the cells of the other. For example, some of the wood-digesting protists in the termite gut themselves contain symbiotic bacteria. Many animals, unicellular and multicellular, contain symbiotic green algae or blue-green Bacteria inside their cells. How the symbiotic organisms got inside the cells is often unknown, but it is characteristic of eukaryote cells to ingest small objects (this is how amoeba feeds, and how our white blood cells deal with bacteria) and ingestion is one plausible source of intracellular symbionts. Another possibility is that the intracellular symbionts were originally parasites, which entered the host cell by their own activity and were eventually put to use by the host.

With mitochondria what was once theory seems now confirmed: they were originally free-living, aerobic Bacteria. In remote Precambrian times, as oxygen accumulated in the atmosphere some simple organism, intolerant of oxygen, avoided oxygen poisoning by acquiring Bacteria that could deal with it more efficiently. By natural selection this association eventually became obligatory to both partners, and was passed on to eukaryotes – the descendants of the original partnership. Confirmation of that theory has come from comparing DNA sequences of mitochondria with those of prokaryotes. Mitochondrial sequences are most closely related to those of a particular group of Bacteria, the alpha subdivision of the purple non-sulphurs, a subgroup of Proteobacteria. Although it has been proposed that mitochondria arose more than once, by associations between different eukaryotes and bacteria, it seems that all the eukaryotic mitochondrial DNAs so far sampled cluster together in the tree, implying that they have a common origin and are descended from one initial symbiont.

The same sort of story was repeated in the origin of plants. All green plant cells contain chloroplasts, packages containing chlorophyll and the enzyme systems responsible for photosynthesis. Plant chloroplasts behave in the same way as mitochondria – they contain their own DNA and RNA, are capable of reproduction, and when the plant cell divides a variety of mechanisms ensure that each daughter cell receives at least one chloroplast. As with mitochondria, chloroplasts appear to be derived from free–living prokaryotes, blue-green Bacteria, that were ingested as symbionts by primitive eukaryotes – the Precambrian ancestors of green plants and other plant-like groups. Comparisons of chloroplast DNA sequences with those of prokaryotes place chloroplasts among the blue–green Bacteria, the Cyanobacteria. As with mitochondria, there has been argument about whether chloroplasts originated once or several times, by symbiotic association between cyanobacteria and different eukaryote hosts. The current belief, based again on comparing DNA sequences, is that all chloroplasts are descended from one original symbiotic association but that

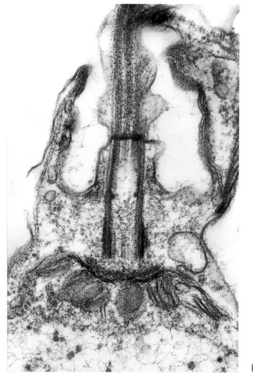

(a)

Figure 15.5

Electron micrographs of sections through cilia and a flagellum

(a) Vertical section through the base of the flagellum of a protist, showing the basal body, where the flagellum is anchored in the cell (x 44 000).
(b) Cilia on the surface of a comb-jelly (ctenophore), a marine animal (x 56 700).
(c) At higher magnification (x 125 000) the nine bundles of marginal tubules and the two central tubules are clearly visible.

(b)

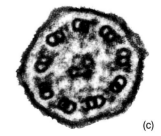

(c)

some eukaryote lineages (algae of various kinds) acquired their chloroplasts at second hand, through ingesting and eventually forming permanent symbiotic associations with single-celled eukaryotes already possessing chloroplasts. Evidence of this secondary symbiosis is found in some single-celled algae that contain chloroplasts enclosed in four membranes, with vestiges of the cytoplasm and nucleus of the original eukaryotic symbiont in the compartment between the inner and outer pair of membranes.

The symbiotic theory of the origin of mitochondria and chloroplasts was proposed long before anything was known about DNA sequences, and has received triumphant support from comparisons of DNA. But that support does not get to the root of the matter: how eukaryotes first originated. The acquisition of mitochondria evidently happened before that of chloroplasts, for mitochondria are more closely integrated into the cell (with many processes controlled from the nucleus) than are chloroplasts, and many single-celled eukaryotes lack chloroplasts and apparently never had them. There are also single-celled eukaryotes that lack mitochondria. These organisms, mostly parasites or inhabitants of environments low in oxygen, were thought to be survivors from the earliest stages of eukaryote evolution, before symbiosis with the bacteria that were to become mitochondria, but it seems more likely that they have lost mitochondria, having no need of them in their special environment.

The symbiosis theory, so successful in explaining the origin of mitochondria and chloroplasts in eukaryotes, has been extended to account for some of the other major differences between prokaryotes and eukaryotes (Section 15.1). These extra symbioses are much more speculative, possibly because the features they explain are older and more closely integrated into the host, so that their history is more obscure. Symbiosis has been invoked to explain mitosis and the presence of flagella and cilia. These are mobile, hair- or whip-like structures projecting from the surface of the cell, which either move the organism (if it is small enough, as in protists and sperm) or move fluids past the cell (as in the ciliated tracts in our lungs and sinuses). Under the electron microscope cilia and flagella have the same structure (Fig. 15.5): when cut across, each shows nine bundles of protein tubules spaced around the margin and two central tubules. These tubules are of constant size, with a diameter of 0.000025 mm. The embedded part of the cilium or flagellum ends in a basal body, a cylindrical structure with nine peripheral bundles of tubules but

without central tubules. These basal bodies are, like the mitochondria and chloroplasts, semi-autonomous. They are able to reproduce (not by dividing but by the development of a new basal body alongside an old one), and newly formed basal bodies can produce cilia or flagella as outgrowths. Basal bodies contain RNA, and there is (disputed) evidence that in some organisms they contain DNA.

In form, size and activity flagella and cilia resemble mobile, thread-like prokaryotes – the spirochaete Bacteria. Various unicellular eukaryotes have a symbiotic relationship with spirochaetes, which attach to and project from the surface of the cell – they were at first mistaken for flagella. It has thus been proposed that all organisms with ciliate or flagellate cells (virtually all eukaryotes) are descended from Precambrian organisms that acquired mobility from a symbiotic relationship with spirochaetes, and in which the relationship later became more intimate and eventually obligatory. However, all attempts to demonstrate molecular homology between the protein tubules in cilia or flagella and proteins in spirochaetes have so far failed.

Other structures in eukaryote cells also have nine bundles of protein tubules like those in cilia, flagella and their basal bodies. The most important of these are the **centrioles**, two (or more) bodies lying near or in the nucleus that are also capable of independent reproduction. The centrioles play an essential role in mitosis because they produce the spindle to which the chromosomes become attached. The spindle is formed of protein tubules homologous with those inside cilia and flagella, and these spindle tubules grow out from the centrioles in the same way as cilia and flagella grow out from basal bodies. The equivalence of basal bodies and centrioles is proved in sperm cells, where the tail flagellum grows out from the centriole, and in many single-celled flagellate organisms, in which the basal body of the flagellum acts as a centriole in mitosis. Mitosis cannot occur in the absence of centrioles, and until recently it was thought that there are some living eukaryotes that have flagella (and basal bodies) but did not have true mitosis and had never evolved it. This implied an earlier stage of eukaryote evolution, before flagellar basal bodies had been recruited or modified as organizers in mitosis. However, it now seems unlikely that there are any surviving eukaryotes without mitosis. The spirochaete symbiosis theory, which accounts both for the organs that make simple eukaryotes motile and for features of mitosis, is an attractive story. But it predicts that flagellar proteins should be homologous with proteins in spirochaetes and, as noted at the end of the last paragraph, all attempts to demonstrate that homology have so far failed. My feeling is that the spirochaete symbiosis story is losing support.

This is a pity, because the spirochaete theory filled out the early stages in a plausible story about events in Precambrian times that produced the eukaryotic cell (Fig. 15.6). Without it we lack a convincing explanation for the origin of flagella, the nucleus and mitosis, all of which are present in the most primitive eukaryotes but absent in all prokaryotes. One theory is that another symbiotic event is involved, a proto-eukaryote lineage having engulfed and formed a permanent association with an archaebacterium (member of Archaea) which became the nucleus (Fig. 15.6). This idea was recently proposed in order to account for similarities between protein-coding genes in eukaryotes and Archaea (Fig. 15.4) and between ribosomal genes in Archaea and Bacteria, suggesting that the protein-coding (nuclear) genes and the translation apparatus (ribosomes) of eukaryotes might have different histories. The theory has hardly been tested, and has nothing to say about the origin of mitosis and flagella.

An alternative scenario is that eukaryotes developed the nucleus, mitosis and flagella without symbiosis. The nucleus could have appeared, burying the cell's DNA in an internal compartment, when early eukaryotes first acquired their ability to engulf food particles. Centrioles and a simple form of mitosis might have developed before or after the nucleus, and flagella would appear later, as outgrowths from centrioles (= basal bodies) formed of the same tubule proteins as those in the mitotic spindle (Fig. 15.6). Unfortunately, this story makes no predictions that I can think of, and so may be untestable. It is interesting that one of the thermophile Archaea, *Thermoplasma*, differs from other prokaryotes in having its DNA insulated or protected by protein, like eukaryotes, and also lacks the rigid cell wall found in almost all prokaryotes. Since the Archaea may be closer to eukaryotes than are Bacteria (Fig. 15.4), *Thermoplasma* (which lives best at 60°C) could give some hint of what a proto-eukaryote was like, but the gap between it and the most primitive living eukaryotes is bridged only by speculation.

15.5 The origin and maintenance of sex

Mendelian genetics, as described in the early parts of this book, are applicable only to organisms that reproduce sexually, having haploid and diploid stages in the life cycle and a reduction division (meiosis) at the transition from diploid

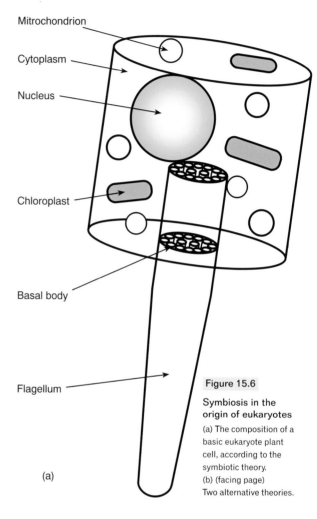

Figure 15.6

Symbiosis in the origin of eukaryotes

(a) The composition of a basic eukaryote plant cell, according to the symbiotic theory.
(b) (facing page) Two alternative theories.

Mitrochondrion

Cytoplasm

Nucleus

Chloroplast

Basal body

Flagellum

(a)

to haploid. This complex sexual cycle does not occur in prokaryotes, and some primitive unicellular eukaryotes (some amoebae and flagellate protists) that lack it and have never evolved it.

Sexual reproduction must first have evolved in simple unicellular eukaryotes. In such organisms the sexual cycle is the exact opposite of asexual reproduction: in asexual individuals the cell divides into two new ones but in sexual reproduction two cells come together and fuse to produce a single new one. We have to ask why such an uneconomical and complicated process should evolve; what are its advantages?

One answer can be given by contrasting the history of new, advantageous, mutations in sexual and in asexual species (Fig. 15.7). Suppose that in an asexual colony of

Theory 1 – symbiosis   (b)   Theory 2 – elaboration of cell membrane

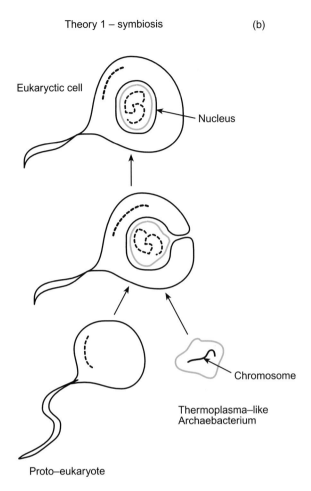

Eukaryctic cell

Nucleus

Chromosome

Thermoplasma–like
Archaebacterium

Proto–eukaryote

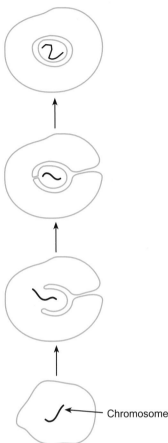

Chromosome

parasitic bacteria a mutation conferring resistance to strep-tomycin appears in one individual, and in another a second mutation gives immunity to penicillin. Each of these muta-tions is valuable, but successive treatment with the two antibiotics will exterminate the population: it could survive only if the two mutations were combined in one individual. This is not possible unless the second mutation were to arise in a direct descendant of the bearer of the first muta-tion. In a sexual species, gametes produced by one mutant or its descendants can fertilize gametes produced by the other, combining the two mutations in one or more mem-bers of the population. But in asexual species only a single mutation (or the descendants of a single individual) can be selected at any one time, and evolution will necessarily be slow. In sexual species every individual is the descendant of two members of the previous generation, and has four

grandparents, eight great-grandparents, and so on, so that many different mutations can combine in one individual and selection can act on many different mutations at the same time.

In asexual reproduction, the descendants of one individ-ual will be genetically identical, generation after genera-tion, unless mutations occur. In sexual reproduction, every individual is a new, unique combination of genes caused by the reshuffling of each parent's chromosomes and genes in meiosis and their combination in fertilization. As a result, sexual species are much more variable and have greater flexibility or plasticity to meet environmental changes.

So one advantage of sex is that it increases variability and the potential rate of evolution. The disadvantages of sex are considerable. In sexual species (except hermaphrodites) about half the population may be thought of as effectively

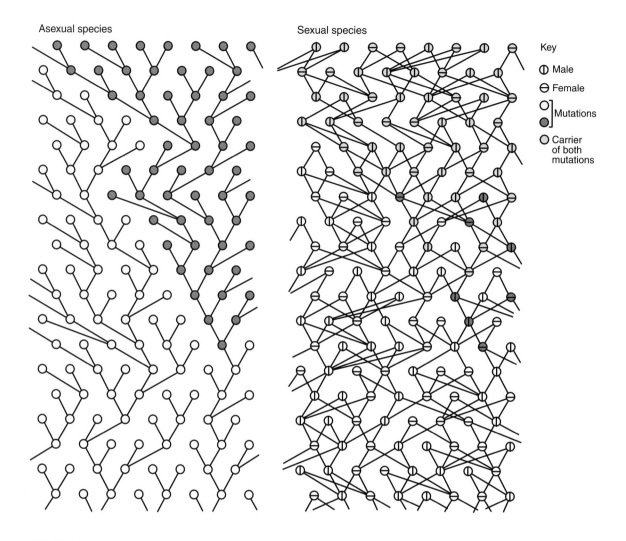

Figure 15.7

Sex can combine advantageous mutations

The effects of natural selection in sexual and asexual organisms. Each circle represents an individual. In the sexual population the two sexes are indicated by a horizontal or a vertical bar. Colours indicate new, advantageous mutations. In the sexual species two new mutations can combine and may spread rapidly through the population. In the asexual species the two mutations are in competition, and one or other will be eliminated.

sterile, because the next generation cannot contain more individuals than those gametes produced by one sex in the parent generation. In most animal species, where the sexes are fully differentiated and males contribute nothing more than one sperm to their offspring, females are effectively suffering a 50% loss of fitness (defined as success in reproduction) compared with asexual females who simply repro-

duce themselves, because half the offspring of sexual females are males, contributing virtually nothing to the following generation. A population of asexual females can, in theory, produce twice as many offspring as an equal population divided into two sexes. The cost of sex, in terms of fitness, is enormous. And simply for sexual reproduction to occur, organisms have adopted extraordinary stratagems and structures. So it is not surprising that many species, descended from sexual ancestors, have given up these stratagems and gone back to the asexual system, often by doing away with males. Asexual reproduction may occur by budding off new individuals, as in many plants and simple animals, or the female may produce eggs that develop without fertilization, either by a modification of meiosis which gives diploid eggs or after 'fertilization' of the egg by another of the haploid nuclei produced in meiosis. This sort of 'virgin birth' occurs in many insects, especially aphids (greenfly), and in many plants, including successful and familiar ones such as brambles and dandelions. Greenfly and dandelions have not given up sex entirely. In greenfly, summer generations are all females, reproducing by virgin birth, but in autumn a generation of males and females is produced; these animals mate and give rise to a new all-female generation ready for next spring. In dandelions the western European races are triploids (see page XX), always reproduce asexually, and have given up sex entirely and irreversibly. But in central Europe there are sexual diploid races which serve as a store of variation and of new asexual triploid races.

If a species or population gives up sex entirely, like the dandelions of western Europe, it may be very successful in the short term, and may well last for millions of years. But it will be genetically uniform and capable only of generating variants within a narrow range. Habitats change with time, however, and such species are more likely to become extinct than their sexual competitors. Evidence of this can be seen in the tree of life, the pattern of relationships between species and groups, in which asexual species or groups of species are concentrated in the terminal twigs, implying that they are short-lived and unsuccessful compared with their sexual relatives.

Returning to the origins of sex, the process must have first evolved in simple unicellular organisms. Today, such organisms behave rather like the greenfly mentioned above; they reproduce asexually for many generations and mate rarely, sometimes only when conditions are deteriorating. In this way they have the benefits of asexual reproduction (efficient increase in numbers when conditions are favourable) and of sex (increased variability to cope with hard times). But because mating is rare in many unicellular organisms it is hard to catch them at it, and we still know far too little about the details of sexual reproduction in these simple forms. This may well be one of the reasons for our almost complete ignorance about how sex originated.

The main theoretical difficulty in speculating about the evolution of sex is that natural selection works on individuals, but the advantage of sex (increasing variability and the rate of evolution) applies to populations – sex increases the variability of the population. Natural selection has no foresight, and cannot work for long-term advantages conferred on the population as a whole: we need a more convincing explanation for the origin of sex.

In primitive unicellular organisms there is no differentiation of sexes and mating involves the whole cell: two cells come together, their cell walls break down at the area of contact, the contents mingle and nuclei are exchanged. There is a resemblance here to the engulfing or eating of one cell by another, as in amoeba, and it is possible that sex, like mitochondria and chloroplasts, originated from this habit of eating other organisms. The sexual cycle in unicellular organisms is very varied. Meiosis does not always precede fertilization, as it does in plants and animals. In some protists it is the zygote, the product of fertilization, that undergoes meiosis, so that the whole life cycle is haploid except for the diploid zygote.

Several possible explanations have been proposed for the origin of sex by natural selection, as an adaptation. Two that are worth mentioning concern the load of deleterious mutations and the arms race between organisms and their parasites.

Parasites provide us with one of the most obvious examples of the struggle for existence. In our own species the ravages of parasitic diseases – malaria, bilharzia, plague, smallpox, even influenza – are as obvious today as in history. Without drugs or immunization, defences against parasites depend on resistance, either developed through the immune system, during infection, or innate (genetic), as in the mechanisms that render us immune to most of the species that parasitize or infect our domestic animals (and vice versa). The 'rate of evolution' argument for the origin of sex is weak because it depends on future advantage, on generating variation to cope with changes in the environment that lie in the future. But parasites are an aspect of the environment, and can be thought of as an environment that is constantly changing as the defences of an organism are probed by a variety of different species or strains of parasite. In this light the success of sex in generating variation might be advantageous in the short term, in the next generation, by providing new combinations of genes conferring resistance or (from the parasite's point of view) overcoming resistance. The argument works for both host and parasite, and computer modelling of simple genetic systems shows that it may work in practice.

Another possible explanation for sex is a version of the argument illustrated in Fig. 15.7, the advantage of sex in combining favourable or advantageous mutations. Deleterious mutations are eliminated by natural selection (just as favourable mutations are propagated by it), as described in Chapter 7, but at a cost – the selective deaths necessary to eliminate their carriers. However, deleterious mutations are far more numerous than favourable mutations, and in asexual lineages they might accumulate faster than they can be eliminated. Imagine an asexual species that reproduces by binary fission (splitting into two daughters) in which each individual suffers, on average, one slightly deleterious mutation during its life. Both daughters inherit the parental mutation and if population numbers are to remain constant one (on average) must die before reproducing (see Fig. 2.1). The survivor acquires a further deleterious mutation, therefore passing two to its descendants, and so on through the generations (Fig. 15.8(a)). Obviously, such a species has no future, for before long the combined effects of the accumulated mutations, however slight their effects, will drive it to extinction. The mutation rate is too high for natural selection to cope. Contrast this situation with a sexually reproducing species suffering the same mutation rate. Because only half the chromosomes of each

parent are passed on to any one descendant, only half the gametes will carry a harmful mutation. In the next generation the two parental mutations (one from each) will be distributed like one of Mendel's factors in peas (see Fig. 4.1) with (on average) a ratio of 1:2:1 – one of four offspring carries both parental mutations, two carry one mutation, and one has none. In this situation, natural selection could cope, by the differential success in each generation of the lucky ones that start with undamaged genes (Fig. 15.8(b)) Mutation rates as high as this may seem unlikely, but in early Precambrian times, when sex originated, DNA was surely not copied so accurately as it is today, after 2000 million years of selection to improve the fidelity of the copying process.

Eliminating harmful mutations by natural selection, by death of their carriers, is a rather wasteful way of repairing damaged DNA. In bacteria, one function of their peculiar sexual processes – passing lengths of DNA from one individual to another – seems to be a more direct method of repairing damaged DNA, analogous to proof-reading one text against another. Meiosis may also have originated for the same reason, for when the homologous chromosomes pair off and exchange DNA (Figs 4.4 and 5.2) there is a sort of proof-reading or comparison of the matching texts.

With the establishment of the sexual cycle in simple unicellular organisms, however it came about, the rate of

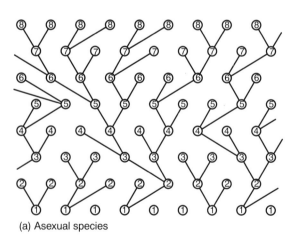

(a) Asexual species

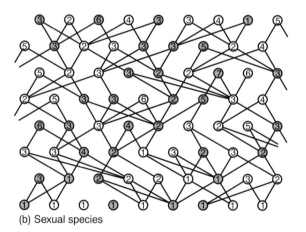

(b) Sexual species

Figure 15.8

Sex can shed harmful mutations

What might happen in (a) asexual and (b) sexual species if each individual acquires one deleterious mutation (the patterns of descent are the same as in Figs 2.1a and 2.1b). The numbers in the circles (symbolizing individuals) are the number of deleterious mutations carried.

evolution must have accelerated. Diverse unicellular organisms have experimented in colony formation, and from such experiments must have come different lines of multicellular organisms, represented today by fungi, plants and animals. The advantage of multicellular organization is that it allows division of labour. Different cells or groups of cells can become digesters, sensors, movers, conductors, and the reproductive function is left to a few cells – those that give rise to gametes. The 'body' of a unicellular organism is potentially immortal, because the reproductive function is not relegated to one part of it. It is only in multicellular organisms, in which reproductive potential is confined to the gametes, that there evolved the burden we all carry – the certainty of death.

16 Evolution and humanity

New books on human evolution come out almost weekly, and many of them are good. This chapter can therefore be brief. I will say something on the when, where and how our species evolved; comment on evolution and human behaviour and say something to meet the reader who has got this far and asks 'how does all this affect my life?'

16.1   Out of Africa?

It is therefore probable that Africa was formerly inhabited by extinct apes closely allied to the gorilla and chimpanzee; and as these two species are now man's nearest allies, it is somewhat more probable that our early progenitors lived on the African continent than elsewhere.

Charles Darwin, *The Descent of Man*, 1871

The traditional approach to the problem of when and where humanity originated has been to search for fossils. A more practical and rigorous approach is through comparing theories of the kind shown in Fig. 13.11. The problem itself – when and where humans first appeared – has meant different things to different people. Humans have been defined as a species that makes tools, or is self-aware, or uses language, or walks fully upright, or whose brain exceeds a certain size. If humans are defined in any one of these ways the question 'where and when?' becomes one of where and when tool-use (or language, or upright posture, etc.) arose. The least ambiguous way of posing the question is through patterns like those in Fig. 16.1: in asking where and when the human lineage originated we are interested in the speciation event that separated us from our closest relatives. If the theory shown in Fig. 16.1(a) or (b) (two versions of Darwin's insight in the epigraph) were correct

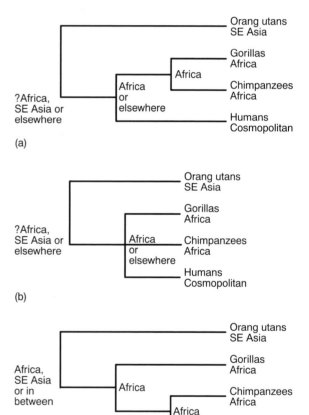

Figure 16.1

Where did humans originate?

Estimates of where the human lineage originated, based on different estimates of the pattern of relationship between apes and humans. (a) and (b) show two patterns considered likely in Darwin's time, (c) the present best estimate.

we might guess that humans arose in Africa, where chimpanzees and gorillas are found; or in south-east Asia, where gibbons and orang utans live; or somewhere in between. 'In between' was the choice of Ernst Haeckel (1834–1919), Darwin's apostle in Germany in the 1860s and 1870s, who placed 'Paradise' in the Indian Ocean and produced an elaborate and detailed plan of human dispersal over the globe (Fig. 16.2). But if the correct theory of our relationships is the one in Fig. 16.1(c), as it seems to be (Section 13.3), then the fossil hunters are right to search in Africa, for both of our closest relatives are found only there, and the two successive speciations shown in Fig. 13.1(c) should have happened in Africa. As for when we separated from the apes

(the more recent of the two speciations), the fossils say at least four million years ago, when man-like fossils (*Australopithecus*) appear in Africa.

There is still controversy over a more recent phase in human evolution, between two competing theories: 'out of Africa' versus 'parallel development' – the belief that modern humans (*Homo sapiens*) originated independently by parallel evolutionary changes in several parts of the world. This can be seen as a version of an ancient controversy, reaching back into the eighteenth century, between 'monogenists', who placed all humans in a single species, and 'polygenists', who placed them in several. Darwin, a monogenist, noted sarcastically in 1871 that the polygenists

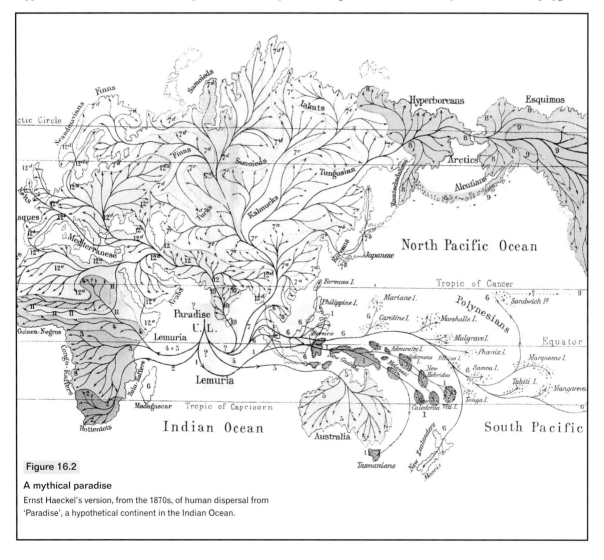

Figure 16.2

A mythical paradise

Ernst Haeckel's version, from the 1870s, of human dispersal from 'Paradise', a hypothetical continent in the Indian Ocean.

were unable to agree on whether humankind represented two, three, four, five, etc. (up to 60) different species (Haeckel favoured 12). Darwin expected that, once evolution was accepted, 'the dispute between the monogenists and the polygenists will die a silent and unobserved death.' But in a different form it is still with us, although no one now disputes that all living humans belong to the same species.

In general, adherents of 'out of Africa' (the modern monogenists) give more weight to evidence from molecular genetics, particularly from analysis of the DNA in mitochondria. Mitochondria (Section 15.1) are organelles within the cell descended from free-living bacteria that integrated into the cells of eukaryotes billions of years ago. Their inheritance is maternal, through the egg, with no contribution from the paternal sperm. Mitochondrial DNA evolves about ten times as fast as nuclear DNA, and so is a valuable tool in reconstructing events in the fairly recent past. Owing to its maternal inheritance, mitochondrial DNA has a pattern of descent like that in asexual species where, purely by chance, all members of a population will be descended from a single individual in some earlier generation (see Fig. 7.3). Mitochondrial DNA sequences can therefore be used both to reconstruct patterns of relationship between human races and, assuming a molecular clock (Sections 9.1 and 9.2), to attempt to date the 'mitochondrial Eve', the maternal common ancestor of our mitochondria*.

A famous study, published in 1989, included 182 variant human mitochondrial DNA sequences, gathered from 241 individuals by the economical method of plucking a couple of hairs from each – the hair bulb contains enough tissue to sample the DNA. Comparison of these DNA sequences with each other and with samples from chimpanzees produced the tree in Fig. 16.3(a), with a cluster of Africans near the root, implying a 'mitochondrial Eve' living in Africa. Estimates of divergence between human and chimpanzee DNA sequences indicated that the time from the present to 'mitochondrial Eve' was about 4% of the time from the present to the separation of humans and chimpanzees – if that separation happened five million years ago 'Eve' lived about 200 000 years ago. This study provoked a lot of justifiable technical criticism, directed both at the

---

*Use of the name 'Eve' does not imply that there was ever a time when only a single woman existed. Mitochondria are inherited like asexual organisms, as in Fig. 7.3(a), but our chromosomes are inherited as in Fig. 7.3(b), and other women (and men) living at the same time as 'Eve' have passed on their chromosomes.

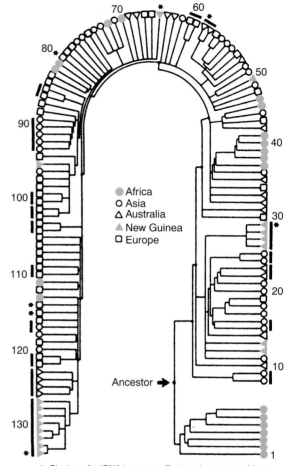

* Clusters of mtDNA types specific to a given geographic area
| mtDNA types found in more than 1 individual

(a)

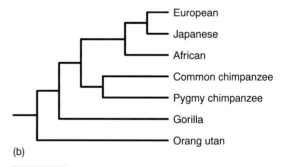

(b)

Figure 16.3

Trees of human mitochondrial DNA sequences

(a) A tree based on 134 samples of human mitochondrial DNA. There is a cluster of Africans near the root.
(Courtesy of *Nature*, 1st January 1987, Rebecca L Cann, Mark Stoneking and Allan C Wilson)
(b) A tree based on complete mitochondrial DNA genomes from apes and humans.

dating of 'Eve' and at the African rooting of the tree. But more recent studies, based on smaller human samples and different segments of mitochondrial DNA, agree in favouring an African 'Eve' and in ratios of the time since 'Eve' and since the human/chimpanzee split of about 1:25 or 1:30, dating 'Eve' at 150 000 – 500 000 years ago. The most accurate estimate comes from comparisons of complete sequences of mitochondrial DNA (about 15 000 bases) from three humans and four apes (Fig. 16.3(b)). This study dated 'Eve' at about 140 000 years ago and the split between the European and Japanese samples at about 70 000 years ago.

The male equivalent of the mitochondrion, which is inherited only through females, is the Y chromosome, passed on only through males (in mammals males are XY, females are XX; page 14). As with mitochondria, all existing human Y chromosomes will be descended, by chance, from one ancestor ('Adam'). Two different point mutations, shared by all sampled non-African Y chromosomes, have recently been identified; the unchanged ancestral base is found only in a proportion of Africans. The implication, as with mitochondria, is that all humans have an African ancestry, and the time of the Y chromosome mutations is estimated at 100 000–200 000 years ago.

These estimates are unacceptable to modern polygenists, who favour the 'parallel development' or multiregional hypothesis and argue that *Homo sapiens* evolved independently in different regions, by parallel changes converting small-brained archaic types into modern humans. Advocates of this theory are impressed by the fossil record, which reveals man-like fossils (*Homo erectus*) scattered through southern Europe and Asia (as well as Africa) at least a million years ago, the fossils in each region resembling its modern inhabitants in certain respects. A 'mitochondrial Eve' in Africa less than half a million years ago is clearly incompatible with this theory, which requires the racial variation found today to have roots extending back a million years or more. The 'out of Africa' theory, supported by both DNA sequence analysis and some interpretations of the fossils, implies that there have been two human migrations out of Africa – an earlier one, a million or more years ago (*Homo erectus*), and a more recent one, well within the last half million years, from which all living populations (*Homo sapiens*) are descended. It is plausible (but untestable at present) that the impetus behind the highly successful second radiation out of Africa was the development of language (Section 16.3).

Figure 16.4

Young and adult chimpanzees

The youth looks like a close cousin, whereas the old male seems to have little to do with us.
(From *Naturwissenschaften,* Volume 14, 1926, pp. 447-448)

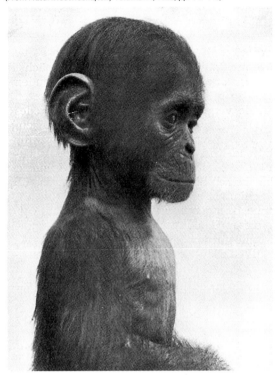

16.2 Baby face

The 'why?' of human origin is answered less easily than the 'where?' and 'when?'. We now know, with real confidence, that our nearest biological relatives are chimpanzees, and fossils tell us that the human lineage was recognizable some four million years ago, in creatures (*Australopithecus*) that differed from apes primarily in bipedality rather than in brain size.

The anatomical differences between modern humans and chimpanzees, though striking to our eyes, are not profound – and when adult humans are compared with young stages (late embryos and juveniles) of chimpanzees or gorillas many of the most striking differences disappear. Our large brains, thin-walled and bulbous skulls, flat faces, upright heads and lack of hair are all found in embryo or baby apes (Fig. 16.4). The differences between chimpanzees and humans could be due to a few, perhaps only two or three, mutations in the regulatory genes that control our development. As a result of these mutations human development is slowed down or retarded – hence the helplessness and huge, lolling head of newborn babies, the late eruption and small size of our teeth, and our very long period of infancy and growth, which allows the learning from parents and others that is so characteristic of humanity. The human gestation period is the same as that of gorillas and only six weeks longer than that of chimpanzees, yet our postnatal period of immaturity, about 18 years, is twice that of these apes – as is our lifespan.

These facts show that human development is retarded relative to the apes. A similar argument can be made when comparing apes with other primates (e.g. monkeys, lemurs). Apes, and humans in particular, are examples of a mode of evolution named **paedomorphosis** – retention of juvenile ancestral features in adult descendants. This type of retarded development seems to have been common in evolution and has been invoked to explain the origin of groups ranging from those containing many thousands of species, such as vertebrates as a whole, down to those with a dozen or so species such as the ratite birds (flightless; ostrich, kiwi, etc.). The classic case is salamanders like the axolotl (*Ambystoma mexicanum*) and its relatives, which mature as overgrown larvae, some of which can be induced to transform into 'adults' by thyroid hormone.

All this does not reduce the gap between us and chimpanzees to nothing. For example, human males have larger penises and human females have larger breasts (whether or not they have given birth) than any ape. Paedomorphosis (or fetalization) will not explain these differences. They are also unlikely to have arisen through natural selection (with the environment as the selective agent), and are best put down to Darwin's alternative – sexual selection, in which the opposite sex is the selective agent. Further differences include the chromosome fusion which gives us 46 chromosomes, whereas chimpanzees and other apes have 48, and the six inversions of parts of chromosomes which differentiate us from chimpanzees (Section 6.2). These chromosome mutations are sufficient, I hope, to frustrate anyone who is brave (or insensitive) enough to persevere with the ultimate experiment of trying to hybridize the two species.

16.3 Language

Darwin realized that languages are related and have evolved in much the same way as species: they may be classified into families and subfamilies (implying more or less recent common ancestry); 'rudimentary organs' no longer serving any useful purpose exist in language just as in the structure of organisms; and languages become extinct (like species) and if they do so never reappear.

The analogy is imperfect, for language may spread by conquest or commerce as well as by descent – the ubiquity of English in North America, Australasia and large parts of Africa and Asia is no indication of immediate common descent among those speaking it. Language may also evolve by horizontal transfer (transfection), by borrowing words from other languages – 'OK' is now supposed to be the most widely used expression in the world (with 'Coca-Cola' second). Nevertheless, languages have split (speciated) and changed, mostly by gradual, roughly clock-like processes like those in non-coding DNA, and these changes are correlated with history. In Britain there are still obvious differences in accent and dialect over a few tens of miles, and differences that may impede understanding over a few hundred. By contrast, in North America obvious regional

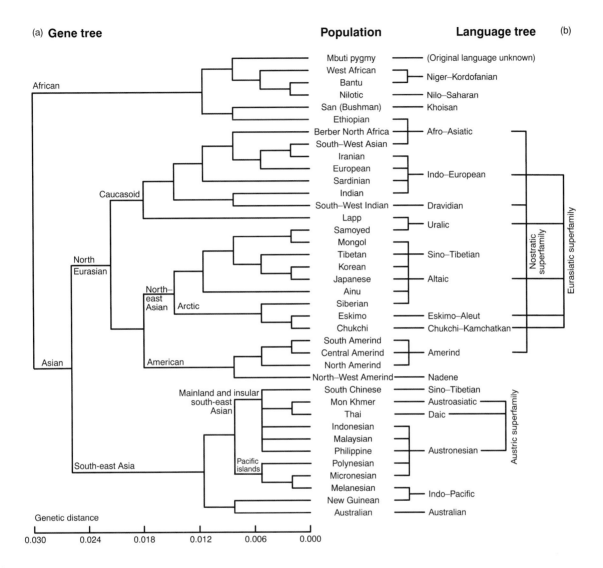

(a) Gene tree

Population

Language tree (b)

Figure 16.5

Trees of human languages and genes

The gene tree (a) estimates historical relationship between human groups, and the language tree.

(b) Relationships between languages. Although the language tree lacks the deeper connections shown in the gene tree, the two are mainly in agreement.

Courtesy of Patterns in Evolution, Roger Lewin, Scientific American Library.

accent and dialect (in English) seem confined to the east coast, particularly to the Atlantic provinces of Canada, New England and the old South, each reflecting the speech patterns of seventeenth-century colonists from different parts of the British Isles.

Thus an evolutionary tree of languages can be made and compared with one based on human genes or bits of DNA (Fig. 16.5). There are about 5000 existing languages, and language certainly evolves much more rapidly than genes. We read Chaucer's fourteenth-century English with difficulty, eleventh-century 'English' is virtually a foreign language, and seventh-century 'English' would have been understood by contemporary speakers of 'Dutch', 'Danish' or 'German'. But despite different rates of change there is generally an excellent match between the gene tree and the language tree – apart from one difference: geneticists are convinced that human genes have one common ancestor but linguists are not yet agreed on whether language arose once or several times. Agreement is unlikely, because languages change at such a rate (one estimate is 20% per 1000 years) that all traces of common ancestry are likely to be eroded in less than 10 000 years. The match between the gene and language trees presumably persists because languages changed more slowly until about 5000 years ago, when empires and armies first began to appear.

As for the origin of language, attempts to educate young chimpanzees have had some success when the animals are taught sign language, which they can reproduce, rather than speech, which they cannot. But it has not yet been shown that chimpanzees are capable of syntax, putting together units (of one or more words) according to rules that allow different or original meanings to be expressed and understood (syntax can exist in sign language, just as in speech or writing). Chimpanzees cannot reproduce speech because their anatomy does not permit it; the tongue and larynx are not up to the job – as in other mammals, their larynx is opposite the internal nostrils, and the two can be brought together so that the animal can feed and breathe at the same time. Human babies have the same arrangement until about three months of age, when the larynx moves deeper into the throat and will no longer engage the internal nostrils (try it). The change in position, disadvantageous because of the hazards of choking on food, is advantageous in giving space for the tongue to mould and modify sounds from the larynx, permitting articulate speech. Descent of the larynx in humans is reflected by a slight flexure of the base of the skull. We know that the larynx is in the 'wrong' (primitive, non-speech) position in chimpanzees, as it appears to be in early members of the human lineage (*Australopithecus*). It seems to be in the 'right' (speech) position in all fossils assigned to *Homo sapiens*, back to over 100 000 years ago. Then there is a 'grey area' covering more than a million years of human history, with fossils showing some flexure of the skull base but less than in modern humans. The consensus among experts is that neither *Homo erectus* nor Neanderthals (*Homo neanderthalensis*, who became extinct about 40 000 years ago) were capable of fully articulate speech. Coupled with the molecular evidence this implies that fully articulate speech was confined to *Homo sapiens*, and to the second human expansion out of Africa which replaced *Homo erectus* and Neanderthals.

At present the favoured theory about human language is one that originated with Noam Chomsky: language is instinctive in *Homo sapiens*, and our brains are 'wired' with the rules of syntax just as the brains of songbirds are programmed for song. Birds do not produce the song characteristic of their species spontaneously, but learn songs and the local dialect from their parents. In the same way, humans do not speak spontaneously but effortlessly learn one (or more) languages from their parents or guardians, usually during the second and third years of life; after that, acquiring a language becomes much more difficult. Although words are learned, the child does not learn grammar (syntax) by example, but is genetically programmed with its basic or universal rules.

Language is obviously advantageous to our species, and so, like any other adaptation, its development should be explicable by natural selection. But language does not fossilize and as yet the gap between the two-word signs of an educated chimpanzee and the sensible chatter of a three-year-old child is bridged by no more than speculation.

16.4 Sociobiology – Evolution and human behaviour

A woman is only a woman, but a good cigar is a smoke.
Rudyard Kipling

Remarks like Kipling's are now deeply unfashionable, reflecting a social climate that we believe has passed away. But there is a subdivision of evolutionary biology named sociobiology in which the theories of evolution, ecology and ethology (study of behaviour) are combined in an attempt to explain and interpret the social life of animals (plants, and the simplest animals, have no recognizable social life). Kiplingesque ideas come close to the surface when sociobiologists interpret human society and the genetic basis of human behaviour.

No biologist now disagrees about our close kinship with apes and, more remotely, with monkeys, other mammals and other vertebrates. Disagreement concerns the extent to which we have cast off the effects of that ancestry. For example, sociobiologists argue that kin selection (Section 8.5) is applicable to human societies; that our altruism is due to 'genes for altruism' and has been shaped by the same selective forces as those which have produced sterile worker castes in social insects. They hold that the concept of 'investment ratio' (discussed in Section 8.5 in connection with social insects) gives a genetic basis for Kipling's male-chauvinist epigraph, or for the rhyme that was the only fruit of the philosopher William James's experience with drugs:

> Higamus hogamus,
> Woman is monogamous.
> Hogamus higamus,
> Man is polygamous.

According to this interpretation women are necessarily tied to their children, and will pass on their genes most effectively by selecting as mates males who will stay with them. Fidelity can be tested, for example, by the woman insisting on a long courtship. Males, on the other hand, can pass on their genes most effectively by promiscuity, for their investment in each child may be minimal. Theories of this sort can explain dowries, male dominance and other unfashionable aspects of society.

Ranged against the sociobiologists are those who emphasize not our evolutionary heritage through kinship with other animals, but our own unique and fast-moving vehicle of change, cultural evolution. In simple animals, and in plants, the only information passed from generation to generation is the information in the genes. Change can come about only by changes in genes, and is necessarily slow. In higher animals, such as mammals, parental care of the young allows more information to be passed on: the young are capable of learning by imitating their parents. In our species this instruction of the young has been taken much further, but in addition we have a unique and highly effective means of transmitting information – language. Human language was at first spoken, and so effective only face-to-face, but with successive inventions of writing, printing, radio, recording and electronic information technology, we can now learn not only from our parents and social group but also from people long dead or on other continents. Cultural evolution shifts evolutionary change from slow trial-and-error Darwinism to a more Lamarckian mode, instruction by the environment rather than selection by it. The pace of cultural evolution has been accelerating wildly over the last century or so, but it has plainly been effective since the invention of writing, and through oral tradition must have been acting since language first developed. Given cultural evolution, the manifest flexibility of the human brain, and the opportunities we have to analyse our own behaviour through introspection, the 'anti-sociobiologists' contend that we have thrown off our brutish heritage, and that our behaviour is dominated not by our genes but by reason, training, tradition, fashion and all the other forces to which we are exposed.

Which side is right? Genetics cannot help much here, for there is not much reliable evidence to set on either side. There is still little evidence that any trait of human behaviour is genetically determined, except in the crudest terms. One example is the eccentric behaviour of sufferers of Huntington's disease, caused by a dominant mutation that affects about 1 in 20 000 births (the socialist singer Woody Guthrie and his mother were victims). Another is the behaviour of people with Down's syndrome (they have an extra copy of chromosome 21; see page 33) and one form of Alzheimer's disease, in which a mutation of a gene on chromosome 21 has the same effect as the extra copy in Down's syndrome, causing accumulation of a protein in the brain that gradually entangles and kills the nerve cells. There have been recent reports of discovery of 'the gene for alcoholism' and 'the gene for schizophrenia', but both claims were withdrawn. On the other side there is little clear evidence that any behavioural trait is *not* genetically determined, except that neither you nor I will accept that our genes dictate the work we do or how we will spend this evening.

As with intelligence (Section 5.2), perhaps the most reliable evidence relating genes and human behaviour comes from studies of identical twins separated soon after birth and reared apart. The most complete study had identified 56 pairs by 1990. Each twin underwent a battery of psychological tests and the general conclusion was that the correlation between the performance of each pair was about the same as that of identical twins who had been reared together. The study produced some remarkable anecdotes – such as two American firemen, both keen on motorcycles, who met and discovered that they were identical twins separated at birth after colleagues noticed the physical resemblance between them; or two 40-year-old twins, also separated at birth, who found on first meeting that they both built miniature furniture in their basement workshops and had each been married twice, first to a Linda and second to a Betty! You might be ready to believe that an interest in firefighting or doll's-house furniture could be programmed by the genes. I assume you will not believe that a Linda–Betty sequence could be so programmed.

The sociobiology debate is an updated version of the 'nature or nurture' controversy of the early days of genetics: do genes or the environment have the upper hand? That debate fizzled out with the answer 'partly' — in some instances the genes win, in others the environment, but there is no clear-cut answer (Section 5.2). No doubt the same is true of genetic and cultural contributions to behaviour. The 'nature or nurture' debate was never merely scientific; it had strong political overtones.

The Lamarckian doctrine of directly inherited environmental effects persisted in Russia until the 1960s and, under Lysenko's leadership, dominated Soviet genetics in the 1930s and 1940s. The reasons were that Lamarckian inheritance offered a short-cut to the perfectibility of man, his crops and domestic animals, that matched the Marxist programme. In the west, on the other hand, the doctrine of Darwin and Mendel — primacy of the genes and the infinitely gradual pace of change — suited capitalist society. For example, Darwin's inheritance allowed him to devote himself to science, for it relieved him of the necessity of earning a living. His servants, too, were born to their station in life. So it is with the sociobiology debate: the left are ranged on the side of cultural evolution, rapid change and the possibility of betterment; and the right are on the side of the genes, and our heritage from the distant past.

16.5 Human nature

I see no good reason why the views given in this volume should shock the religious feelings of any one.
Charles Darwin, *The Origin of Species*

This concluding section is to meet the reader who has got this far and asks 'how does all this affect my life?' Evolution theory has been called 'what ultimately may prove to be the greatest revolution in the history of thought'. It has changed, and is changing, our understanding of our history,

purpose and future. Thousands of pages have been written on these lines, and I have no capacity to add much to them. Evolution theory in the crude form of survival of the fittest — or 'social Darwinism' as it was politely called — has been used to justify Hitler's racial policies, grinding the faces of the poor, and many other examples of man's inhumanity to man. Has it anything more positive to offer? Some people find it profoundly depressing. They read genetics and evolution as proof that we are the products of blind chance, nothing but pointless experiments in protein chemistry. My reply is that I find it no more depressing to be a chemical experiment than an experiment in ethics, which is the Christian message read from the same nihilist or 'nothing but' viewpoint.

It is true that evolution theory is no substitute for religion, though some still try to make it one. But it does contain a message about our relationship to the rest of nature that is more positive than the Old Testament message that man is unique, made in the image of God, and that the rest of creation is there for him to exploit. It will take us a long time completely to lose that attitude, although its results are daily more obtrusive. The message of evolution is that we are not unique. We are animals, members of the same lineage as the wood louse and the shrew. We are social mammals, and as such our purpose is not merely to reproduce, yet our racial pride or patriotism and our social striving may be no more or less commendable than the tribal loyalty of baboons. Our uniqueness lies in our brains, tongues and hands, which have allowed us to accumulate knowledge (or to decrease ignorance), building on the experience of previous generations in a way that no other species can. It is consciousness of the past and anticipation of the future, derived from this knowledge, that give us control over the beasts of the field. And if we are alone in the universe we are alone in the sense that we are our own masters and have no one to blame but ourselves, for if control over the future is anywhere it is in our own heads.

17 Who's who

17.1 Before Darwin

Figure 17.1

Carl Linnaeus (1707–1778), Swedish naturalist, Professor of Botany at Uppsala

Linnaeus laid the foundation of systematic biology with his method of naming and classifying animals and plants (page 100). Modern botanical nomenclature dates from 1753 (Linnaeus's *Species Plantarum*) and zoological nomenclature from 1758 (Linnaeus's *Systema Naturae*, 10th edition). Linnaeus was no evolutionist, but his system of classification is readily adapted to evolutionary interpretation (see Fig. 13.2). This lithograph is from a portrait of Linnaeus in old age (1775). (Courtesy of the Linnean Society)

Figure 17.2

Lamarck (Jean-Baptiste-Pierre-Antoine de Monet, chevalier de Lamarck, 1744–1829), French naturalist

Lamarck was the first to publish a reasoned theory of evolution (or transformism, as it was then called), in his *Philosophie Zoologique* (1809). Lamarck is unfortunate to be remembered today chiefly in connection with the inheritance of acquired characters (page 52). In France he is regarded as the father of evolution, yet he died in poverty and was vilified by the scientific establishment of his time, especially by Baron Cuvier, the greatest authority in zoology and a believer in the theory that the history of life consists of repeated creations and catastrophic extinctions. Darwin, too, was unnecessarily contemptuous of Lamarck, for he also upheld the inheritance of acquired characters. This engraving shows Lamarck in 1821. (Courtesy of the Linnean Society)

Figure 17.3

Robert Chambers (1802–1871), Scottish writer, publisher and amateur geologist

With his brother he founded the firm of W. & R. Chambers. In 1844 the first book on evolution in English, *Vestiges of the Natural History of Creation*, was published anonymously in London. In 1884, after his death and 40 years of rumour, Chambers was acknowledged as the author of this book. Though denounced and condemned by scientific and religious authorities alike, Chambers' book went through ten editions in ten years, and Darwin later acknowledged that it had helped to clear the ground for his own work. Chambers' thoughts may have been turned towards hereditary change (evolution) by the fact that he and his brother were born with six digits on both hands and feet. The extra fingers and toes were removed in childhood, but the operation left Robert lame. This engraving shows him aged about 60.

17.2 Darwin

Figure 17.4

Figure 17.5

Charles Robert Darwin (1809–1882)

After training for medicine and the church, Darwin educated himself in science as naturalist on the surveying voyage of HMS *Beagle* (1831–1836). He began a notebook on transmutation of species in 1837, shortly after returning to England, and in 1838 hit on the idea of natural selection as an explanation of evolutionary change. With a prosperous background (his father was a successful doctor, and both his mother and his wife were Wedgwoods), Darwin worked as an amateur at his home, Down House, near Bromley, Kent. Although limited by chronic ill health to three or four hours' work a day, he laboured for twenty years compiling material for a huge manuscript on evolution by natural selection. This work might never have gone to the printer in his lifetime, but in 1858 he received a manuscript from A. R. Wallace (see right), then in the Molluccas, which showed that Wallace had independently arrived at exactly the same theory as Darwin. Darwin and Wallace published a short joint paper in 1858, and Darwin immediately set to work abstracting his large manuscript. The result was *On the Origin of Species* (1859). Darwin published eight more books during his lifetime, the most important of which are *The Variation of Animals and Plants under Domestication* (1868), *The Descent of Man, and Selection in Relation to Sex* (1871) and *The Expression of the Emotions in Man and Animals* (1872). I recommend his autobiography, written for his children, and only published posthumously. Down House is open to the public as a museum of Darwiniana. Figure 17.4 shows Darwin in about 1857, just before he began *On the Origin of Species* (courtesy of the Linnean Society), and Figure 17.5 shows him in 1881 in 'his usual out-of-doors dress'.

17.3 Darwin's friends and supporters

Figure 17.6

Thomas Henry Huxley (1825–1895), zoologist and educator

After he qualified as a doctor Huxley's career was moulded, like Darwin's, during a Royal Navy surveying voyage, on HMS *Rattlesnake* (1846–1850). He was Professor at the Royal School of Mines in London and later at the Royal College of Surgeons. When *On the Origin of Species* was published Huxley was an immediate convert to Darwin's views ('How extremely stupid not to have thought of that,' he said) and he became Darwin's chief supporter in England, notably at the 1860 meeting of the British Association in Oxford, when he crossed swords with the Bishop of Oxford, and in several battles with Richard Owen (page 152). Huxley was a brilliant talker, writer and thinker and some of his mental qualities come across in this photograph, taken in 1857. (Courtesy of the Linnean Society)

Figure 17.7

Alfred Russel Wallace (1823–1913), traveller and naturalist

Wallace travelled and collected on the Amazon (1848–1852) and in the Indo-Australian archipelago (1854–1862), where he independently came to the same theory as Darwin, and spurred Darwin into writing *On the Origin of Species*. In later years Wallace published a major work on the distribution of animals and books on travel, natural selection and evolution. He was never convinced that evolution could explain the human brain, and was an ardent socialist, spiritualist and opponent of vaccination – something of a crank. He was also a modest, kindly and most humane and civilized person, as this 1895 photograph suggests. (Courtesy of the Linnean Society)

Figure 17.8

Reverend John Stevens Henslow (1796–1861), Professor of Botany at Cambridge 1827–1861

When Darwin was an undergraduate at Cambridge, reading for the church, Henslow befriended him and was responsible for his appointment to HMS *Beagle*. He was a lifelong correspondent with Darwin, and was chairman of the 1860 British Association meeting at which T. H. Huxley took up the cudgels on Darwin's behalf. This photograph was taken in about 1856. (Courtesy of the Linnean Society)

Figure 17.10

Asa Gray (1810–1888), American botanist, Professor of Botany at Harvard 1842–1888

Gray was Darwin's chief supporter in America. Although he met Darwin in London in 1839 and again in 1850, the two did not begin to correspond until 1855 (the photograph shows Gray at about this time). An 1857 letter from Darwin to Gray, summarizing Darwin's theory, formed part of the 1858 Darwin/Wallace paper. In the years following publication of *On the Origin of Species*, Gray devoted himself to propagating Darwin's ideas in America, and overcoming the influence of his Harvard colleague Louis Agassiz (page 152). (Courtesy of the Linnean Society)

Figure 17.9

Sir Charles Lyell (1797–1875), geologist

His book *Principles of Geology* (1830–1833) supported uniformitarianism, the doctrine that the distant past is to be explained only by the forces that we see in operation today. These ideas were directly opposed to the catastrophism of Cuvier and others, and were an inspiration to Darwin's thinking during the voyage of the Beagle. Lyell became a close friend of Darwin, and it was to him that Darwin appealed for advice when he received Wallace's manuscript in 1858. The photograph shows him in about 1856. (Courtesy of the Linnean Society)

Figure 17.11

Sir Joseph Dalton Hooker (1817–1911), botanist

Hooker's father, Sir William, also a botanist, was the first Director of Kew Gardens. Joseph Hooker, like Darwin and Huxley, began his scientific career on a Royal Navy ship, HMS *Erebus*, which cruised the Southern Ocean in 1839–1843. Hooker founded the science of plant geography and was the first person to whom Darwin showed the outline of his theory (1844). With Lyell, Hooker arranged publication of the Darwin/Wallace paper in 1858. Hooker was married to Henslow's (Fig. 17.8) daughter, and succeeded his father as Director of Kew Gardens in 1865. He is photographed in the same year as Lyell. (Courtesy of the Linnean Society)

17.4 Darwin's opponents

Figure 17.12

Louis Agassiz (1807–1873), zoologist and teacher

Agassiz was born in Switzerland and studied in Germany, with the nature-philosophers. Like Owen, he was a protégé of Cuvier, and to the end of his life maintained Cuvier's views on repeated creations of species and catastrophic extinctions. After a brilliant career in Europe (amongst his achievements was recognition of the Ice Age) he went to America in 1846, and in 1848 became professor at Harvard, where he founded the Museum of Comparative Zoology. Agassiz gained a towering reputation in America by his lecture tours, writing and magnetic personality. He opposed Darwin's theory with all his powers, right up to his death. The photograph shows him at Harvard in about 1860.

Sir Richard Owen (1804–1892), comparative anatomist

Owen belongs to an entirely different tradition from Darwin. He had studied in Paris under Cuvier and was influenced by the ideals of the German nature-philosophers (his concepts of homology and archetype, shown in Fig. 13.1, derive from the German school). When *On the Origin of Species* was published he wrote a scathing anonymous review but in his scientific work he wrote of evolution in obscure passages that manage to suggest both disagreement with Darwin and that he had thought of it first. From 1856 to 1883 he was Superintendent of Natural History at the British Museum, and at the age of 75 was responsible for moving the collections into the new building at South Kensington, which opened in 1881. Figure 17.13 shows him in about 1855, his hand resting on the crocodile skull shown in Fig. 13.1; Figure 17.14 shows him in old age, about 1885. (Courtesy of the Linnean Society)

Figure 17.13

Figure 17.14

17.5 Geneticists

Figure 17.15

Johann Gregor Mendel (1822–1884)

Mendel was born in Austria of peasant stock and entered the Augustinian monastery at Brno in 1843. The monastery was a centre of learning and research, and Mendel was instructed in agriculture and botany and sent to Vienna University. He was teaching at the Technical School in Brno during the period (1856–1864) when he carried out his experiments with peas (page 10) in the monastery garden. In 1868 he became Abbot of the monastery where he continued work on hybridization of plants and of bees. He died unrecognized, and achieved immortality in this world only in 1900, when his work on peas was rediscovered and his theory confirmed. The engraving shows him in about 1860.

Neo-Darwinism, combining Darwinian theory with Mendelian genetics, was developed in the 1920s and 1930s mainly by the work of three scientists – Sir Ronald Fisher, J. B. S. Haldene and Sewall Wright (Figs 17.16, 17.17, 17.18).

Sir Ronald Fisher (1890–1962)

Fisher was an English mathematician who worked at Rothamsted and Cambridge. He published a series of papers (starting in 1918) analysing consequences of Mendelian inheritance, and in 1930 his book *The Genetical Theory of Natural Selection* laid the foundations for reconciling genetics with natural selection. (Courtesy of Joan Fisher Box)

Figure 17.16

Figure 17.17

J.B.S. Haldane
(1892–1964)

Haldane was a polymath whose works included visions of the future (*Daedalus, or Science and the Future*, 1924; *Possible Worlds*, 1927), a fine childrens' book (*My Friend Mr Leakey*, 1937) and *The Causes of Evolution* (1932), a book on genetics and Darwinism that was almost as influential as Fisher's and far more readable. He taught at Cambridge (1922–1932) and in London (1933–1957), and emigrated to India in 1957, in disgust at the British goverment's doings in Suez.

Figure 17.18

James Watson (left) and Francis Crick in 1953, with their original model of the DNA molecule in the Cavendish Laboratory, Cambridge

Both are still active in molecular genetics. Crick (born 1916) remained at the Medical Research Council's molecular biology laboratory in Cambridge until 1977, when he moved to the Salk Institute in San Diego, California. Watson (born 1928) was Professor of Biology at Harvard and is now at the Cold Spring Harbor Laboratory in New York State. Watson and Crick shared the 1962 Nobel Prize for medicine. (Photo A.C. Barrington Brown and Weidenfeld & Nicolson Archives)

Figure 17.19

Sewall Wright
(1889–1988)

Sewall Wright's long life (he died at 98) spanned the entire history of genetics, from the rediscovery of Mendel's work in 1900 to molecular genetics of the 1980s. Wright was born in Massachusetts, grew up in Illinois, worked on guinea-pig breeding for the US Department of Agriculture (1915–1925) and then taught (and continued to study guinea-pigs) in Chicago (1926–1955) and in Madison, Wisconsin (1955 into old age). Like Fisher and Haldane, during the 1920s he applied a gift for mathematics to combining genetics with natural selection, but he gave greater attention than them to random effects, particularly in small populations (genetic drift). After retiring from Chicago, Wright published the four-volume treatise *Evolution and the Genetics of Populations* (1968–1978). The photograph shows Wright in Chicago in 1954; there is a legend (untrue) that the guinea-pig in his hand was afterwards used to wipe the blackboard. (Courtesy of Stephen Lewellyn, Lewellyn Studios, Chicago)

Figure 17.20

Motoo Kimura
(1924–1994), Japanese geneticist

Kimura is father of the neutral theory of molecular evolution (Chapter 9), which he first proposed in 1968 and elaborated in a 1983 book. His early work, including study for a PhD in the United States, was devoted to developing improved mathematical analyses of genetic drift, the process of neutral or non-selective change proposed by Sewall Wright (above). The discovery of the structure and role of DNA in the 1950s enabled Kimura to understand and predict the differences between Darwinian evolution, of phenotypes exposed to natural selection, and non-Darwinian evolution, of DNA which may be invisible and so immune to natural selection. Kimura's early ideas received support during the 1980s, as DNA sequences became available. (Courtesy of Genetics Society of America)

Further reading

Within only the last ten years, hundreds of books and thousands of scientific papers on evolution have been published. Most of this is written for specialists and assumes a background knowledge of biology. All I provide here is a selection of the better and more recent books, emphasizing those that do not demand too much prior knowledge in the reader. Paperback editions are cited where they exist; almost everything cited contains a bibliography. Like all scientific fields, evolution theory is constantly changing, and this week's idea may be proved wrong next week. So the most difficult thing is to keep up to date. The monthly journals *Scientific American* and *Natural History* (the latter published by the American Museum of Natural History) are available in most libraries. *Scientific American* contains three or four articles a year on the latest ideas in evolution, written for the non-specialist and with illustrations of very high calibre. Since 1974 *Natural History* has contained a monthly column, 'This View of Life' by Stephen J. Gould, which reviews developments in evolution theory in brilliant style. Eight collections of Gould's essays are published by Penguin as *Ever Since Darwin* (1980), *The Panda's Thumb* (1983), *Hen's Teeth and Horse's Toes* (1984), *The Flamingo's Smile* (1987), *An Urchin in the Storm* (1990), *Bully for Brontosaurus* (1992), *Eight Little Piggies* (1994), and *Dinosaur in a Haystack* (1997).

General works

The two standard modern textbooks on evolution are:

Evolutionary Biology (2nd edn 1986),
Douglas J. Futuyma. Sinauer: Sunderland, MA.

Evolution (2nd edn 1996),
Mark Ridley. Blackwell: Oxford and Boston.

Encyclopaedic, and excellent value, is the 1000-page text produced by the Open University:

Evolution: a Biological and Palaeontological Perspective (1993),
Peter Skelton (ed.). Addison-Wesley: Wokingham, UK and Reading, MA.

On the Origin of Species, Charles Darwin (1859)
The first edition is now priced in thousands. There were six editions during Darwin's lifetime, and many later; one in Penguin is easily available. Well worth reading or dipping

into, for it is still full of interest. It was written for the general reader, and though Darwin was not proud of his prose or his mind, both come over strongly. There are several compendia of extracts from this and Darwin's other books. After *On the Origin of Species*, Darwin's most important book is:
The Descent of Man, and Selection in Relation to Sex (1871 and many later editions).

Two excellent short books on Darwin are:

Darwin (1982),
Wilma George. Fontana Modern Masters: London

Darwin (1982),
Jonathan Howard. Oxford University Press Past Masters: Oxford

The best biographies of Darwin are:

Darwin (1991, 1992)
Adrian Desmond and James Moore. Michael Joseph: London, 1991; Penguin: London, 1992.

Charles Darwin Vol. 1 *Voyaging*.
Janet Browne, Jonathan Cape: London.
Vol. 2 to follow

The leading modern evolutionary theorist is John Maynard Smith. His easiest (and earliest) book is:

The Theory of Evolution (1st edn 1958; 2nd edn 1963; 3rd edn 1975)
Penguin: London; reissued 1993 by Cambridge University Press: Cambridge.
A text written at semi-popular level.

Evolutionary Genetics (1st edn 1989; 2nd edn 1998),
Oxford University Press: Oxford.
A fairly difficult text.

Did Darwin Get it Right? (1993),
Penguin: London.
A collection of essays on 'games, sex and evolution'.

The Major Transitions in Evolution (1995),
John Maynard Smith and Eörs Szathmary. W.H. Freeman: Oxford and New York.
This is not easy reading but is a brilliant survey of major events from the origin of life to the origin of language.

Richard Dawkins writes about evolution at a less technical (or more militant and sensational) level than Maynard Smith. His books include:

The Selfish Gene (2nd edn 1989),
Oxford University Press: Oxford.

The Blind Watchmaker (1988),
Penguin: London.

River Out of Eden (1995),
Phoenix: London.

Climbing Mount Improbable (1997),
Penguin: London.

Books on particular topics

This list is in alphabetical order by author. Numbers in parentheses after each entry are the chapters or sections of this book to which it is most relevant.

The Book of Man (1994),
Walter Bodmer and Robin McKie. Abacus: London.
A popular account of many aspects of modern molecular biology and human genetics. (4-6, 16)

The Rise and Fall of the Third Chimpanzee (1991),
Jared M. Diamond. Vintage: London.
The third chimpanzee is humanity; Diamond emphasizes our closeness to chimpanzees and draws conclusions about many aspects of our past and future. Particularly strong on interpreting human sexuality and on the damage we are inflicting on the environment. (16)

The Language of the Genes (1994),
Steve Jones. Flamingo: London.

In the Blood: God, Genes and Destiny (1996),
Steve Jones. Harper Collins: London.
Excellent and lively accounts of (mainly) human evolution and genetics. (4-6, 9, 10, 16)

The Eighth Day of Creation 1979, reissued 1995), Horace Freeland Judson. Penguin: London
A superb account of the early development of molecular biology. (4, 9, 10)

The Neutral Theory of Molecular Evolution (1983),
Motoo Kimura. Cambridge University Press: Cambridge.
Kimura's account of his theory; technical, but very clearly written. (9, 10)

The Structure of Scientific Revolutions (2nd edn 1970),
Thomas S. Kuhn. University of Chicago Press: Chicago. (14)

The Doctrine of DNA: Biology as Ideology (1993),
Richard C. Lewontin, Penguin: London.
Lewontin is a distinguished geneticist and evolutionist, and also a Marxist. His book is a reasoned argument against sociobiology and tyranny of the genes. (16)

Fundamentals of Molecular Evolution (1991),
Wen-Hsiung Li and Dan Graur. Sinauer: Sunderland, MA.
A good short textbook; technical. (5-10)

Patterns in Evolution: The New Molecular View (1997),
Roger Lewin. Scientific American Library: New York.
Nicely illustrated semi-popular account of molecular evolution. (9, 10, 15, 16)

Popper (1972),
Bryan Magee. Fontana Modern Masters (paperback): London.
Easy introduction to the philosophy of science. (14)

Symbiosis in Cell Evolution (1981),
Lynn Margulis. W.H. Freeman: San Francisco.

Origins of Sex: Three Billion Years of Recombination (1986),
Lynn Margulis and Dorion Sagan. Yale University Press: New Haven.
Two books by Margulis (one written with her son), the leading advocate of symbiosis in the origin and early evolution of eukaryotes. (15)

The Language Instinct (1995),
Steven Pinker. Allen Lane The Penguin Press: London. (16)

Unended Quest (1976),
Karl Popper. Fontana (paperback): London.
Popper's intellectual autobiography. (15)

The Fossil Trail (1995),
Ian Tattersall. Oxford University Press: Oxford and New York.
Popular account of fossil evidence for human evolution. (16)

The Double Helix (1968),
James D. Watson. Penguin: London.
The inside story of the discovery of the structure of DNA. (4)

The Beak of the Finch (1994),
Jonathan Weiner. Alfred A. Knopf: New York.
The story of Peter Grant and his family's long-term study of Darwin's finches on the Galapagos. (8, 11, 12)

Glossary

Abbreviations: adj. adjective; Gr. Greek; L. Latin; n. noun; v. verb.

Words in *italics* are defined elsewhere in the glossary.

Adaptation, n., adaptive, adj. (L. *adaptare*, to fit to). A characteristic of an organism which fits it for a particular environment; the process in which an organism is modified towards greater fitness for its environment.

Adaptive radiation. Evolutionary divergence of a group of related species into different environments or ways of life.

Aerobic, adj. (Gr. *aer*, air; *bios*, life). Requiring free oxygen to live.

Allele, n. (Gr. *allelon*, one another). Any of the alternative states of a *gene*. Alleles occupy the same relative position on the *chromosome*, can mutate to one another, and pair in *meiosis*.

Amino acid, n. One of a large group of organic compounds in which the *molecule* contains an amino group (NH_2) and a carboxyl group (COOH); a subunit of a *protein molecule*.

Anaerobic, adj. Not requiring free oxygen to live.

Archetype, n. (Gr. model). A basic plan or general structure common to a group of *species*; a hypothetical ancestral type.

Asexual reproduction. Reproduction without sex, involving only one parent.

Bacterium (plural bacteria), n. (Gr. *bakterion*, a little rod). A member of an extensive group of minute, unicellular *prokaryote* organisms. Many live as parasites.

Balancing selection. *Natural selection* favouring *heterozygotes* for a particular gene, maintaining *polymorphism* for two or more *alleles*.

Base, n. In chemistry, substances which neutralize acids (the precise meaning is more extensive, but too intricate to explain simply). The four bases in DNA – adenine, guanine, cytosine and thyamine – are used to code for amino acids in groups of three (triplets). See also **Base pair**.

Base pair, n. A term used for the cross-links or 'treads' in the double spiral of the *DNA* molecule, formed by an adenine–thymine bond or a guanine–cytosine bond. A base pair is the 'atom' of heredity.

Biological species. A term for *species* viewed as interbreeding communities, isolated from other such communities by barriers which prevent *gene flow*.

Centriole, n. A small body, found in or near the *nucleus* of each cell, which produces the spindle in nuclear division.

Chloroplast, n. (Gr. *chloros*, green; *plastos*, shaped). Self-duplicating bodies found in the *cytoplasm* of plant cells. They contain chlorophyll, and are the site of *photosynthesis*.

Chromatid, n. (Gr. *chroma*, colour; *idion*, peculiar). One of the two filaments comprising a *chromosome* during nuclear division.

Chromoneme, n. (Gr. *nema*, thread). The single loop of *DNA* found in *prokaryote* organisms; equivalent to the *chromosomes* of *eukaryotes*.

Chromosome, n. (Gr. *soma*, body). Thread-like bodies, comprising *DNA* and protein. There are several in the nucleus of every cell in animals and plants.

Chromosome mutation – see *mutant*.

Cilia (singular cilium), n. (L. eyelash). Fur-like mobile filaments on the surface of cells.

Classical pseudogene – see *pseudogene*

Coding DNA. That portion of an individual's *DNA* that is translated into *protein* (the *exons* of the *genes*).

Concerted evolution. Maintenance of similarity between members of a *gene family* in a *species* or lineage. The genes in the family change over time, but do not diverge from one another.

Conjugation. The process by which bacteria 'mate', and exchange DNA from the bacterial chromosome.

Continental drift. The theory that the continents are not fixed but are in motion relative to each other, drifting, colliding or splitting over long periods of time (see also *tectonics*).

Crossing-over, n. The process in the reduction division (*meiosis*) of reproductive cells in which pairs of *homologous chromosomes* exchange parts.

Cytochrome, n. (Gr. *cytos*, cell; *chroma*, colour). One of a group of iron-containing *enzymes*, found in most animal and plant cells, which play an essential part in oxygen respiration.

Cytoplasm, n. (Gr. *plasma*, shape, body). The living contents of a cell, except for the *nucleus*.

Degenerate, degeneracy (fourfold, threefold, twofold). The number of different *nucleotides* that may occupy a position in coding *DNA* without altering the coded *amino acid*. Degeneracy is a consequence of redundancy in the genetic code (see Fig. 4.7).

Deletion mutation. Loss of part of a chromosome.

Deoxyribonucleic acid, DNA. A substance present in every cell (within the *nucleus* in *eukaryotes*) which gives the hereditary characteristics. See page 19 for its structure.

Diploid, adj. (Gr. *diplos*, double). Having two sets of *homologous chromosomes* in the nucleus; having double the number of *chromosomes* present in the sperm or egg-cells of the *species*.

Directional selection. *Natural selection* that changes the frequency of an *allele*, towards elimination or *fixation*.

Dominance, n., **dominant**, adj. A *gene* or character which is manifested when present in only one *chromosome* of a pair (a single dose, inherited from one parent only). Antonym of *recessive*.

Duplication mutation. Doubling or repetition of part of a *chromosome*.

Ecology, n., **ecological**, adj. (Gr. *oikos*, house; *logos*, reason, speech). The branch of biology dealing with the interactions between *organisms* and their environments.

Ecological (habitat) isolation. Reproductive isolation caused by differences between species in ecological niche or habitat occupied.

Ecological niche. The ecological milieu occupied by a species.

Enzyme, n. (Gr. *en*, in; *zyme*, leaven). Biological catalysts; a class of *proteins* that speed up chemical reactions.

Eukaryote (also spelt eucaryote), n. (Gr. *eu*, true; *karyon*, kernel). A member of the animal, plant, fungus or *protist* kingdoms; an *organism* having *nuclei* and *chromosomes* in its cells. Other organisms are *prokaryotes*.

Exon, n. A piece of coding *DNA*; a portion of a *gene* that is translated into part of the corresponding *protein*.

Fitness, n. Biological success or ability to cope (see page 38 for a technical definition).

Fixation. An *allele* or *mutation* is fixed when it occurs in every member of a population; fixation renders the population homozygous for the allele.

Flagellum (plural flagella), n. (L. a little whip). A whip-like, mobile filament on the surface of a cell.

Fusion translocation, n. A mutation in which non-*homologous chromosomes* are joined end-to-end, so reducing chromosome number.

Frequency, n. Abundance, commonness, number of occurrences.

Frequency-dependent selection. Increased *fitness* of a *genotype* due to its rarity.

Gamete, n. (Gr. wife, husband). A sex cell or germ cell, carrying a *haploid* set of *chromosomes*, whose function is to fuse with another gamete to produce a new individual.

Gene, n. (Gr. *genos*, race). The unit of heredity and development, a stretch of *DNA* (a small piece of a *chromosome*) which contains information necessary for *transcription* into *RNA* and *translation* into *protein*.

Gene conversion. The process invoked to explain *concerted evolution*; unequal *crossing-over* which renders stretches of *DNA* identical.

Gene duplication. Production of two copies of a gene or stretch of *DNA* through unequal *crossing-over*.

Gene family. A number of *genes*, usually but not necessarily on the same *chromosome*, that are related by common ancestry through *gene duplication*.

Gene flow. Flow of *genes* through and between populations, over successive generations, caused by mating and migration.

Gene frequency. An estimate of the abundance of a particular *gene* in a population.

Gene pool, n. The total variety of *genes* and *alleles* present in a *species*: interbreeding may produce any combination of these.

Genetic code. The code by which *base-pairs* in DNA or RNA are translated into amino acids (see page 20).

Genetics, n. The study of heredity.

Genetic drift. Evolutionary change produced by random effects, not by *natural selection*.

Genetic engineering. Insertion (by humans) of genes from one *organism* into the *genome* of another.

Genome, n. In *diploid* organisms, the *DNA* of a complete *haploid* or half set of chromosomes; in *prokaryotes*, the DNA.

Genotype, n. The genetic constitution of an individual; all these *genes* are available for transmission to the offspring of that individual, but not all are manifested in its *phenotype*.

Genus, n. (plural genera), **generic**, adj. (L. birth). A category bearing a name such as *Homo* or *Rhododendron*, and containing one or more *species* which are each other's closest relatives. One or more genera are included in a family.

Geographic isolation. Genetic isolation due to geography. This is not really a kind of 'reproductive isolation', which is normally thought of as depending on differences intrinsic to the organisms. Geographical isolation of a population is often thought to be effective in allowing genetic divergence to take place.

Habitat isolation – see *ecological isolation*.

Haemoglobin, n. (Gr. *haima*, blood; L. *globus*, a ball). A red pigment, a *protein* containing iron which readily combines with oxygen and carries it to the tissues. Found in the blood of many animals.

Haplo-diploidy. The genetic system found in ants, bees, wasps and some other insects, where males are *haploid* and females *diploid*.

Haploid, adj. (Gr. *haplos*, single). Having a single set of *chromosomes*, half the *diploid* number, the chromosome complement of *gametes*.

Hermaphrodite, n. & adj. (Gr. mythology). Having both male and female sex organs, as in many flowers and lower animals.

Heterozygote, n., **heterozygous**, adj. (Gr. *heteros*, other; *zygon*, yoke). An *organism* which will not breed true for a particular character, having received different *alleles* of a *gene* from each of its parents; cf. *homozygote*.

Homeobox, n & adj. A name for a *gene family* coding *proteins* involved in laying down the basic geometry of animals and plants.

Homeosis, n. (Gr. *homoiosis*, becoming alike). *Mutations* in *homeobox genes* (or other processes) that cause one part of an animal or plant to take on the character of another (e.g. insect eye replaced by leg or antenna).

Homology, n., **homologous**, adj. (Gr. *homologos*, agreement). Similarity due to common ancestry; homologous features or homologues agree in relative position and fundamental structure, but may differ in form and function.

Homologous chromosomes. A pair of *chromosomes*, one received from each parent, which have the same series of *genes*.

Homozygote, n., **homozygous**, adj. (Gr. *homos*, same). An organism which breeds true for a particular character, having identical *genes* in a given portion of a pair of *homologous chromosomes*.

Incomplete dominance. The situation where the *heterozygote* for two alleles is intermediate between the two *homozygotes*; the condition arises when both *alleles* are manifested in the *phenotype*.

Insertion mutation. A form of *point mutation* in which one or more extra *base pairs* are inserted into a portion of *DNA*.

Intron, n. A non-coding part of a *gene*, spliced out after *transcription* and before *translation*; cf. *exon* (introns sit between and separate exons).

Inversion mutation. Reversal of part of a *chromosome*, so that the *genes* of that part are in inverse order.

Invertebrate, n. A multicellular animal without a backbone.

Kin selection. Selection of traits that increase the *fitness* of the family rather than the individual.

Linkage group. A number of *genes* or characters which are usually inherited together, because they are neighbours on a *chromosome*.

Macromutation, n. (Gr. *makros*, great). A pronounced genetic change, producing an *organism* which is very different from its parent or parents.

Meiosis, n, **meiotic**, adj. (Gr. diminution). The form of nuclear division in which the number of *chromosomes* is halved. Meiosis occurs in *gamete* formation, when a *diploid* cell produces four *haploid* gametes. The diploid number is restored when two gametes (egg, sperm) fuse.

Melanism, n., **melanic**, adj. (Gr. *melanos*, black). Darkening or blackening of an *organism* by accumulation of black pigment (melanin).

Messenger RNA. A form of *RNA* which is synthesized in the *nucleus*, each *molecule* carrying the message of one *gene*, and is *translated* in the *cytoplasm*, in a *ribosome*.

Metabolism, n., **metabolic**, adj. (Gr. *metabole*, change). The chemical processes through which an *organism* uses food or energy.

Minisatellite DNA. Stretches of *non-coding* human *DNA* that consist of multiple repeats of short (ca 15 *base pair*) sequences; the number of repeats is used in DNA fingerprinting or profiling.

Mitochondrion (plural mitochondria), n. (Gr. *mitos*. thread; *chondros*, granule). Minute bodies present in *eukaryote* cells; they are self-replicating and contain *enzymes* which metabolize oxygen.

Mitosis, n., **mitotic**, adj. (Gr. *mitos*, thread). The process of nuclear division in *eukaryote* cells.

Molecular clock. The rate at which *mutations* accumulate in a *gene* or stretch of *DNA* in a lineage; the theory that mutations accumulate at a roughly constant rate, mostly by *neutral evolution*.

Molecular genetics. Study of heredity at the molecular level, in chemical rather than biological terms.

Molecule, n. (diminutive of L. *moles*, mass). The smallest particle of a chemical substance which can exist separately; molecules are composed of atoms.

Monera, n., **moneran**, adj. (Gr. moneres, *single*). The kingdom containing *prokaryote organisms*, such as bacteria, which have no *nucleus*.

Multiple alleles. Alternative forms of the same part of a *chromosome*.

Mutant, n. & adj., **mutation**, n. (L. *mutare*, to change). A *gene* or *organism* which has undergone a heritable change. *Point mutations* are molecular changes within a *gene*; *chromosome mutations* are rearrangements of *chromosomes*.

Mutation rate. The average frequency with which a particular *mutation* turns up in a population.

Natural selection. The term proposed by Darwin for 'the struggle for existence' or 'survival of the fittest'; differential mortality and reproductive success.

Nucleic acids. Complex organic compounds found in all cells, *DNA* in the *chromoneme* or *chromosome*, and *RNA* in the *nucleus* and *cytoplasm*.

Nucleotide, n. A base in *DNA* or *RNA* (adenine, cytosine, guanine, thymine, uracil).

Nucleus (plural nuclei), n. (L. a small nut). A more or less spherical body found in *eukaryote* cells: it stains deeply with dyes, and contains the *chromosomes*.

Neutral evolution. Evolution by fixation of *mutations* that are neutral or nearly so, through *genetic drift* rather than *natural selection*.

Non-coding DNA. *DNA* that is not *translated*; cf. coding DNA.

Organic, adj. In chemistry, compounds containing carbon.

Organism, n. (L. *organum*, an engine). A living individual.

Phenotype, n. (Gr. *phainein*, to appear). The appearance or characteristics of an *organism* (usually with respect to a particular feature or group of features), the result of interaction between the *genotype* and environment.

Photosynthesis, n. (Gr. *photos*, light). The characteristic process of plant life, in which energy from light is used to produce sugars; chlorophyll is the catalyst in these reactions.

Phylogeny, n., **phylogenetic**, adj. (Gr. *phylon*, race; *genesis*, origin). Study of the evolutionary history and relationships of *species*.

Phylum (plural phyla), n. The largest subdivisions of the plant and animal kingdoms (e.g. Mollusca, Vertebrata).

Plasmid. Self-replicating extrachromosomal DNA molecule in bacteria. Often carries genes for resistance to antibiotics, and can be transmitted to other individuals and even other species.

Plate tectonics. See *tectonics*.

Point mutation. See *mutant*.

Polymorphism, n., **polymorphic**, adj. (Gr. *polys*, many; *morphe*, form). Occurrence of two or more genetically distinct forms of a *species* in the same place.

Polyploid, n. & adj., **polyploidy**, n. (Gr. *polys*, many). An *organism* or *species* with three or more *haploid* sets of chromosomes in each *nucleus*. Triploids have three sets, tetraploids four, hexaploids six, and so on.

Population genetics. The study of heredity in populations rather than individuals – it is conducted by extrapolating from experiments on individuals, and so is largely theoretical and mathematical.

Positive selection – see *directional selection*

Preadaptation. The theory that complicated features may have evolved through stages in which they were adapted to quite different functions from those they now have.

Processed pseudogene – see *pseudogene*

Prokaryote (also spelt procaryote), n. (Gr. *pro*, before; *karyon*, kernel). A micro-organism which has no *nucleus*, such as a *bacterium*. Equivalent to *Monera*; cf. *eukaryote*.

Protein, n. (Gr. *proteios*, primary). A group of *organic* compounds characteristic of living *organisms*; they are chain-like molecules built up from *amino acids*.

Protista (plural protists), n. (Gr. *protistos*, first of all). The kingdom of unicellular *eukaryote organisms*, including plant- and animal-like forms.

Pseudogene. A classical pseudogene is a gene that has been rendered functionless by one or more *mutations* that prevent *transcription* and *translation*. A processed pseudogene is derived from the *RNA transcript* of a *gene*, inserted somewhere in the *genome*.

Purifying selection – see *stabilizing selection*.

Quantum speciation. The theory that some *species* originate in a few generations, by rapid rather than gradual change.

Recessive, adj. The quality of a *gene* or character which is manifested only when present in both *chromosomes* of a pair (a double dose, inherited from both parents). Antonym of *dominant*.

Replacement mutation – see *synonymous mutation*.

Ribonucleic acid, RNA. Substance present in all *organisms* in three forms, messenger RNA, ribosomal RNA and transfer RNA, which function in *translating* the genetic message into *proteins*.

Ribosome, n. (Gr. *soma*, body). Minute particles in the *cytoplasm* of all cells, the site of *protein* synthesis. They consist of protein and ribosomal *RNA*.

Seasonal isolation. Reproductive isolation caused by differences between species in mating season.

Selection coefficient. A number giving an estimate of the relative *fitness* of a particular *genotype* in a particular environment.

Silent mutation – see *synonymous mutation*.

Speciation, n. The production of new *species*, by splitting or division of ancestral species.

Species (plural species), n. (L. kind). A division of a *genus*; a group of interbreeding natural populations which do not interbreed with other such groups.

Spirochaete, n. (Gr. *speira*, a coil; *chaite*, hair). A type of *bacterium*, flexible and spiral-shaped.

Stabilizing selection. *Natural selection* which tends to maintain the status quo, eliminating deviant or abnormal individuals or *genotypes* from a *species*.

Subspecies, n. A division of a *species*; a group of populations which is geographically defined and whose members differ in some way from other subdivisions of that species.

Substitution mutation. A form of *point mutation* in which one *base pair* in *DNA* is changed to another.

Symbiosis, n., **symbiotic**, adj. (Gr. *syn*, together; *bios*, life). An association between members of two or more different *species*, to their mutual advantage. The members of the partnership are symbionts.

Synonymous (or silent) **mutation**. A *point mutation* at a degenerate site, having no effect on the *amino acid* specified. Replacement mutations alter the amino acid.

Tectonics, n. (Gr. *tekton*, builder). Structural geology. Plate tectonics is the theory that the earth's crust consists of a number of plates which are in motion relative to one another.

Tetraploid – see *polyploid*.

Transcription. The first stage in translating the genetic message, when an *RNA molecule* is synthesised on a *DNA* template.

Transduction. The process by which bacteria receive and integrate DNA from another bacterial individual due to transmission by a bacteriophage or virus.

Transfection. Horizontal transfer of *DNA* between members of the same or different *species*, by processes other than descent.

Transfer RNA. A family of *RNA molecules*, found in the *cytoplasm*, which function in *protein* synthesis by attaching specific *amino acids* to the growing protein molecule.

Translation. The second stage in translating the genetic message, in a *ribosome*, where an RNA molecule provides the sequence of *amino acids* in a *protein*.

Translocation mutation. Transfer or exchange of parts between non-*homologous chromosomes*.

Triploid – see *polyploid*.

Vertebrate, n. & adj. An animal with a backbone.

Wild-type, adj. In genetics, the normal or usual condition of an *organism* or a *gene*.

Zygote, n. (Gr. *zygon*, yoke). A fertilized egg; a *diploid* cell produced by fusion of two *haploid gametes*.

Index

Note: *passim* means that discussion of a topic is not continuous but scattered throughout the page ranges mentioned.

The **Glossary** (pp.156-160) has not been indexed as it is self explanatory. The glossary terms in the text have been indexed and the bold text entries clearly refer the reader to the **Glossary**.